T0330182

ROBOTS AND IMMIGRANTS
Who Is Stealing Jobs?

Kostas Maronitis and Denny Pencheva

BRISTOL
UNIVERSITY
PRESS

First published in Great Britain in 2022 by

Bristol University Press
University of Bristol
1–9 Old Park Hill
Bristol
BS2 8BB
UK
t: +44 (0)117 374 6645
e: bup-info@bristol.ac.uk

Details of international sales and distribution partners are available at bristoluniversitypress.co.uk

British Library Cataloguing in Publication Data
A catalogue record for this book is available from the British Library

ISBN 978-1-5292-1271-6 hardcover
ISBN 978-1-5292-1273-0 ePub
ISBN 978-1-5292-1274-7 ePdf

Cover design: Namkwan Cho
Front cover image: Shutterstock/Anton.F
Bristol University Press uses environmentally responsible print partners.
Printed and bound in Great Britain by CPI Group (UK) Ltd, Croydon, CR0 4YY

To my partner, Alev.

To my husband, Tim.

Contents

Acknowledgements

We would like to thank our commissioning editor, Paul Stevens, and editorial assistant, Emma Cook, at Bristol University Press for their enthusiasm in the book and for their ongoing kindness and support throughout the uneasy road to publication. We would also like to thank the anonymous reviewers for their encouraging comments, useful insights and recommendations.

Introduction: Stealing Jobs

Who steals jobs? Who owns jobs? The book unpacks and critically engages with the trope of stealing jobs which has become increasingly important in political and public debates and was a key argument in the campaigns leading to Brexit. The book argues that there is a mutually constitutive relationship between discourses of automation and immigration, which legitimises and entrenches a divisive type of neoliberal governmentality that seeks to weaponise everyday anxieties about job (in)security and the impact of migration on job prospects and public services. This argument is based on the premise that the British labour market is a specific manifestation of neoliberalism that creates a social environment in which people find themselves unable to adapt to the ever-changing demands of the free market and therefore cannot challenge the policies and ideologies of the free market from their position of perpetual insecurity. Thus, neoliberalism demonstrates its agility, as well as ability to create new social relations and to facilitate the formation of new political subjectivities. Therefore, neoliberalism transcends any attempts at reductively defining it as being merely an economic system; rather, it is better understood as an all-encompassing political system (Foucault, 2004).

However, while the quintessential nemesis to the Foucauldian Homo Oeconomicus was traditionally seen as being outside of the neoliberal system, we put forward the proposition that there are *two others* to the Homo Oeconomicus and both of these *others* are to be found within the neoliberal paradigm: these are migrant workers (xeno Homo Oeconomicus) and robots (labour automation technologies and policies). On the one hand, the lens of race and ethnicity has not been sufficiently employed in the conceptualisation of the self–other relations within the neoliberal paradigm despite evidence that neoliberalism facilitates us–them antagonism, as well as fragmentation along ethnic lines (Pencheva and Maronitis, 2018). This analytical silence is significant vis-à-vis the salience of anti-migrant sentiments and arguments in UK political and public debates. This is why we have argued that the other to the Homo Oeconomicus is to be found within neoliberalism itself and that ethnicity is a well-suited lens to explore the tensions that arise between the British and the xeno Homo Oeconomicus.

On the other hand, automation of production and labour are wedded to ideas of growth and productivity, the latter being indispensable to the neoliberal paradigm, which suggests the possibility of understanding robots and machines as another nemesis to the Homo Oeconomicus. Even though automation rapidly transforms the labour market by rendering many jobs obsolete (self-checkouts in supermarkets, self-check-ins at airports, driverless transport vehicles and so on), political and public discussions about labour displacement by automation are not marked by the same level of hostility as discussions about the impact of migrants on the economy and public services. Therefore, the book tasks itself with the comparative discursive exploration of the points of convergence and divergence between the ways migrants and robots are discussed vis-à-vis the political subjectivity of the British Homo Oeconomicus.

It is, however, imperative to note that the book does not consider either immigration or automation as particularly progressive in and of themselves. We are not choosing between them through a cost–benefit analysis in order to arrive at the optimal long-term solution to the peculiarities of the British economy. Instead, we focus on the relationship between automation and immigration because this is a relationship of complex interdependencies and political trade-offs. Aspects of this relationship affect all of us every day: from job insecurity, professional retraining and unemployment, to crippling anxieties about how productive and competitive we are. In other words, the book doesn't task itself with evaluating whether automation or immigration are providing better/worse productivity, but how their relationship is significant and how both are used as tools for managing and controlling the population while entrenching the mantras of neoliberalism.

Ultimately, fears of automation of labour or of immigrants stealing jobs come down to crippling anxieties and insecurities about survival and poverty, that stem from job losses. Thus, the relationship between automation and immigration blurs the line between the political right and left. *The Economist* declared the dividing lines between left and right obsolete as new ones emerged between open and close, between the globalists and protectionists, liberals and conservatives. Policymakers, politicians and political thinkers who embrace openness identify automation as the major challenge of our times whereas protectionists and nationalists discuss immigration as a threat to national economy and culture.

In other words, the relationship between immigration and automation is about optimism and anxiety: optimism about a brighter and progressive future and anxiety about labour displacement. As early as in 1930, John Maynard Keynes considered the vast technological changes in the domains of production and employment and outlined the positive yet somewhat problematic impact of technological progress. It was only a matter of time, Keynes argued in his famous essay 'Economic possibilities for our grandchildren', that the industrialisation of manufacturing, mining and transport would be implemented into agriculture and food production. As soon as industrialisation colonises every aspect of the economy, advanced countries will be able to solve one of the great problems of humankind – that of sustenance. However, Keynes warns us that technological innovation and industrialisation will cause a rather peculiar problem that in the future will dominate public, political and economic debates, namely the disease of 'technological unemployment' (Keynes, 2015: 80). Keynes approaches the apparent problem of technological unemployment in a positive manner and explains it as the direct result of 'our discovery of means of economising the use of labour' (Keynes, 2015: 80). Such a problem, Keynes insists, would be temporal and in the long run the living standards of advanced industrial nations would be four to eight times higher. Even though Keynes understood the existential issues arising out of 'technological unemployment' and in particular out of the newfound freedom that a society of leisure could afford to its members, he did not anticipate how the advent of an automated economy can be halted by immigration and at the same time how it can be used as an instrument for governing the population.

The dynamic relationship between employment and unemployment, security and insecurity, optimism and anxiety, becomes an integral part of the art of neoliberal governance (Maronitis, 2019). This art of neoliberal governance involves the precarisation of individuals and their subsequent relations and hierarchy among them. In the absence of social protection precarisation needs to be regulated in order to remain politically legitimate and more importantly its 'extent must not pass a certain threshold such that it seriously endangers the existing order: in particular it must not lead to insurrection' (Lorey, 2015: 2). Managing this threshold is what makes the art of governing today.

Why Britain?

Our focus on the British economy and society derives from a series of defining contradictions and complexities. First, the British economy has a long and complex history of relying on non-domestic labour, initially through the British empire and subsequently via the UK's (conditional) integration into the structures and practices of the European Union (EU) (Hansen, 2000; Rolfe, 2017; Clayton and Firth, 2018). Following the result of the 2016 referendum on the UK's membership of the EU, the country has found itself in unchartered waters. The UK's current position outside of the EU common market but without robust and properly functioning post-Brexit immigration rules, has given rise to demands for sustaining the current rates of immigrant labour. More specifically, these demands focus on the need for low-waged and/or low-skilled immigrant labour in the service and agricultural sectors of the economy that would generate a competitive and growing economy. Second, despite historically being at the forefront of the Industrial Revolution, contemporary Britain is substantially lagging behind other wealthy liberal democracies and has the lowest levels of automation among the G10 countries (Dellot and Wallace-Stephens, 2017; BEIS, 2019). Therein lie two important contradictions: (1) a structural demand for non-domestic labour versus a general dislike of foreign workers; and (2) a utopic vision of progress through automation and technology versus the prohibitively costly and patchy reality of its rollout.

Third, neoliberalism is not a new phenomenon in Britain, but a deeply entrenched and commonsensical ideology. The vast majority of the literature on neoliberal reforms such as the dismantling of unions, the privatisation of public services and the limited provision of welfare follows a familiar narrative: neoliberalism in all its social, political and economic manifestations is presented as a logical solution to the problematic legacy of centralisation and state control, and as an irreversible force sweeping away the old crumbling practices of individuals and institutions. This narrative is made predictable by the assumed 'newness' of neoliberalism and the inability or unwillingness of the state to catch up with market forces. The austerity programmes and the liberalisation of the labour markets imposed by the International Monetary Fund (IMF), the European Central Bank (ECB), and the European Commission towards Ireland and southern EU member states, such as Italy, Greece, Portugal and Spain, are indicative examples of how neoliberalism has been presented and applied as a new mentality and as a new set of policies and reforms for rectifying the mistakes of state-controlled economies and for liberating individuals from organised labour (Ladi, 2014; Walter, 2016).

However, the British economy and society force us to view neoliberalism in a different light. For more than 40 years fiscal consolidation, privatisation of key industries and services and the marketisation of every aspect of everyday life have become the norm. There is nothing new about neoliberalism. Instead, we are experiencing an old and crumbling system of thought that has been constantly adapting to new economic and political conditions. Since the Brexit referendum and the vote to leave the EU, neoliberalism or, in other words, the free market, low corporation tax, free movement of labour, have been perceived as antithetical to a new set of priorities, namely sovereignty and recognition of national identity. Sovereignty as the guiding principle for exiting the EU, the renewed interest in national identity, and the recognition of the 'White working class' as the authentic voice of and for the nation require an economic and political model centred on patriotic values and on an overall sense of social and economic protection. Furthermore, the partial suspension of economic activity and the re-evaluation of both work and workspace during the COVID-19 pandemic have made clear that neoliberalism is not and should not be an uncontested political and economic mentality.

Therefore, Brexit Britain is the focal point of our analysis. The year 2016 was the year of the Brexit referendum, which legitimised the idea that immigration from the EU and elsewhere is problematic as well as harmful to the economy and society. It was also the year that political parties started to reorient their policies and envision a society and a labour landscape beyond the free movement of labour from and to EU member states. Technology, robotics, automation and artificial intelligence (AI) have been discussed as potential solutions to the lack of immigrant workers as well as to low productivity and stagnant economic growth. Yet, these discussions generate fears and anxieties quite like the ones generated by immigration. What will happen to existing jobs? Will they be replaced by robots? What do workers have to do in order to be prepared for the imminent technological takeover? What are the responsibilities of the state towards those whose professions are most threatened by automation? Can we imagine communities where work is not the central organising principle? Fears of a technological takeover have been mitigated by an impressive body of scholarly and policy work that envisions and advocates a world without work where human associations and activities are shaped by a common sense of creativity and the common good (Dardot and Laval, 2018; Lawrence and Laybourn-Langton, 2018). For example, the popularity of Universal Basic Income (UBI) and its potential to not only support people in a fully automated economy but also to liberate them from the constraints of waged labour and competition, has been growing exponentially.

Brexit might be our starting point but not our year zero. The result of the referendum and its immediate aftermath with regard to national, class and other collective identities, immigration and employment policies constitute a manifestation, but not the conclusion, of long political and social and economic processes, conflicts and contradictions. A referendum that is both an unfinished event and the product of long processes allows the construction of a narrative that goes back and forth: from New Labour's endorsement of the free movement of labour within the EU to the salary thresholds of the Conservatives' immigration policy; from the economic liberalisation of the early 1980s to the alleged protectionism and salary support schemes of the 2020s thus far dominated by the COVID-19 pandemic; from the derision of the working class and the wider discourse on benefits scroungers to the nostalgia of a racialised working class embraced by all political parties; from the contemporary precarious workers to the heroic entrepreneurs of the economics text books; from the fear of immigrants and immigration up until the Brexit referendum in 2016 to lack of immigrant workers in agriculture, hospitality and social care; from the liberal fascination with automation and robotics to the nostalgia for Victorian ingenuity and the slogan of Build Back Better.

Such a narrative is informed and set in motion by an ongoing crisis that appears in different political, social and economic forms. Many of the themes explored in this book entered the respective popular and political lexicons during the 2008 financial crisis and the punitive austerity that followed it. In turn, the central idea and content of the book were conceived and formulated during the political crisis generated by the Brexit negotiations. Finally, most of the book was written during the social and public health crisis caused by the COVID-19 pandemic. The consecutive crises Britain has experienced over the years are neither a simple indication of bad fortune nor a spectacular deviation from normality. Rather, as Dardot and Laval (2019: 19) point out, it is a 'veritable method of government'. Since Britain's 1978–1979 'winter of discontent' and the exhaustion of Keynesianism as capitalism's last-minute saviour, the difficult times and structural problems identified by successive prime ministers and policymakers have been deployed as the stalking horse for 'realistic', 'pragmatic' and above all audacious politics situated within the economic, political and social limits of neoliberalism.

What do you do? Where are you from?

These questions are all too common in the British context. At first glance, these questions seem innocent and mundane enough and most of us have had to answer them, however begrudgingly. However, these questions are profoundly performative as well.

So, *what do you do?* Implicitly or explicitly this question is raising the importance of work for one's identity, wealth, education, cultural disposition and standing in the world. Even though most of us have been asked this question, we are very much aware of how insecure work has become, and how the notion of a steady, well-paid and secure job has become a thing of the past. Yet, we still cling on to the idea that work is a constitutive component for the formation of our identity and the organisation of community. But not all jobs are equal and not all people do the jobs they wish they could do. What is considered to be a good or a bad job depends on wages, the quality of the working environment and conditions, the rights and mobility that a job, or to be more precise the employer, affords you.

The second question of our first conversation is centred on place, origins and belonging. *Where do you come from?* In a similar fashion to the first question, where do you come from seeks answers regarding identity and rights. This is a question that oscillates between the legal and the personal. It is not a question limited to small talk but rather extends to job and visa applications, recruitment procedures, immigration policies and public attitudes. Where do you come from means do you have the right to be here? If so, do you have the right to do the same job as the rest of the people living and working in the country? Do you have the right to the same rights, working conditions and access to welfare as the rest of the workers? Our second question aspires to and partially succeeds in creating a sense of order. In the first instance, it wishes to make it clear that not all workers are equal and should not be equal in the labour market. In the second instance, the question 'where do you come from' indicates that this order is fragile and requires a lot of effort to maintain. A clear legal, social and political distinction must be maintained between 'us' and 'them', between those who have the right and those who do not have the right to take part in the labour market, and those who have the right to welfare and those who do not.

Asking and answering these questions within the context of neoliberalism means that this conversation is taking place between the domains of politics and the economy. We need to consider the fact that developments in the economy and politics are contemporaneous and condition each other. A general convention in this conversation between politics and the economy is to examine the impact policies have on productivity, economic growth and working conditions. But we also need to take into account another, often neglected, dimension of this conversation: the impact of the free market on the direction and relevance of politics. Even though politics and the economy develop contemporaneously and condition each other, they respond to different demands. The former is concerned with growth and maximisation of profits and the latter with the protection of citizens and the redistribution of wealth. At the heart of this conversation are two unresolved issues created by the advent of neoliberalism. First, neoliberalism upheld the liberal idea of organising a society around the principles of the market economy but also advocated for the active role of the state in constructing and maintaining a viable competitive order. Such an order could not possibly exist if corporations and all kinds of private actors were left unsupported and unsupervised by the state's infrastructure and legal frameworks.

Second, another element in this conversation between the economy and politics revolves around technology and more specifically the way technological progress is presented as an irreversible force that demands training and adaptability. The former Prime Minister Tony Blair (2021), in one of his numerous interventions into global politics and the affairs of the Labour Party, argued that the new technological revolution encapsulated in automation, robotics and AI divides the world into those skilled enough to adapt and those who are stuck in an economic and political past of steady and secure labour (Blair, 2021). According to this logic, the inevitable consequences of insecurity and instability that technological developments bring along can only be alleviated by a state-driven educational policy that facilitates and responds to technological change and the demands of the competitive market.

Despite the contemporary significance and relevance of automation and robotics in the domains of employment, productivity, economic growth and status, debates among policymakers concerning the ideological disposition of technological developments remain absent. In key speeches post-Brexit and during the COVID-19 pandemic policymakers avoided the topic of automation and instead focused on the contribution of migrant workers to the economy and how the future flow of migrant labour will depend on the needs of the economy, skills and salary thresholds. Concerns over the future flow of immigrant labour have been expressed by employers and business federations, all of whom demanded continued access to a cheap and flexible workforce without expensive legal complications related to visas and work permits. Everybody wanted a skilled, obedient and highly productive workforce. They wanted *men machines*.

Indeed, the anticipated negative economic impact of regulating and ultimately decreasing immigration could be somewhat mitigated by a series of political, economic and technological adjustments. Policymakers and think tanks expect that over time the competitive labour market will eventually adjust to changes in labour supply through wage increases, digitisation and automation. More specifically, the UK government is planning to substitute labour for labour – that means encouraging British citizens to do jobs currently performed by immigrant workers, and by substituting capital for labour in the form of robotics and automation technologies.

Neoliberalism and Homo Oeconomicus: dead or alive?

The British economy, politics and society are frequently analysed through the prism of neoliberalism and the dominance of free market ideology. Notwithstanding the multiple and contested meanings of neoliberalism, it is generally meant to be understood as an idea originated in the 1930s, advocated by right-wing economists and think tanks, and since the late 1970s largely implemented by politicians across the political parliamentary spectrum. Before we engage with these multiple and contested meanings in a theoretical and empirical manner, we need to locate neoliberalism in the economic, political and social fields which both shapes and is being shaped them. While *Robots and Immigrants* gravitates towards a Foucauldian analysis due to our focus on the national and xeno Homo Oeconmicus as contemporary manifestations of neoliberalism in Britain, the selected empirical material and case studies force us to engage with the wider historical, intellectual and political aspects and analyses of neoliberalism.

In the first instance, neoliberalism exists as an intellectual movement aiming at the revival of liberal thought with an autonomous free market at its core. From the Walter Lippman Colloquium (1938) to the Mont Pélerin Society (1947), and the Chicago School led by Milton Friedman, neoliberalism has been associated with the desire to elevate the free market as the central organising principle of the economy and society. Consequently, analysis of neoliberalism has moved to the domain of politics and policymaking. In the work of David Harvey (2005), Callison and Manfredi, 2019) and Amable and Palombarini (2021), neoliberalism is presented and analysed as a response to over-regulated economies, rigid labour markets and to low productivity. Debates on liberty, regulation and productivity have effectively led to the analysis of subjectivity and of the neoliberal subject. Drawing on Michel Foucault's lectures at the Collège de France in the 1970s, and later on Wendy Brown and Dardot and Laval, neoliberalism is not a concept, or a grand narrative limited to economics and policy but more importantly it expands to the way we conduct ourselves and relate to others. According to Foucault, neoliberalism is a rationality that lacks ideological coherence and inevitably escapes the confines of a top-down affair involving think tanks, policymakers and the general population. Neoliberalism and its associated meanings such as competition, the free market and liberty create the social and cultural conditions for the formation and subsequent classification of subjects.

The foregoing discussion on the semantic evolution of neoliberalism has raised concerns about the intellectual coherence of the concept. Indeed, some have gone as far as to argue that neoliberalism was dead. However, history has taught us that such proclamations are premature and at times naive. Neoliberalism has been declared dead several times in the past only to be resuscitated by economic crises and lack of political imagination. In 1998 Eric Hobsbawm put together a convincing case for the dead ends and subsequent death of neoliberalism. By referring to a series of political and economic crises in the former USSR and in South and East Asia respectively, Hobsbawm argued that the assumption of the free market as a great liberating force is destructive and plays havoc with people's lives. However, Hobsbawm's death of neoliberalism is not a one-off event but rather a process. Neoliberalism might be dead but as Hobsbawm pointed out in 1998, it remained alive and well in the policies of New Labour.

Hobsbawm offers four different yet interconnected explanations for the complex phenomenon of neoliberalism's death and endurance. First, the 'mixed economy' model reached its limits by the end of the 1970s and national governments ideologically inclined to social democracy were unable to control the market forces in their interior. Second, the belief in the self-regulating market propagated by economists like Hayek and Friedman gained traction as the only viable solution for combating inflation and stagnant economic growth. Third, globalisation and the free market, as Tony Blair never stops reminding us, is an inevitability that we all have to deal with, and we need to be politically and economically equipped to harness their benefits. Fourth, New Labour assumed that general elections are won by appealing to the middle classes who have benefited from Thatcher's housing reforms and deregulation of banking.

But why would governments and political parties endorse and apply such a destructive system? According to Hobsbawm (1998) the relationship developed between the economic orthodoxy of neoliberalism and politics is not a relationship sustained by solutions and positive results, but a relationship sustained by the abdication of responsibility considering economic affairs and the direction of the market. Neoliberalism enables a politically justified admission that the economy is beyond the reach of governments and the only responsibility of the latter is to create an appropriate environment for the much desired macro-economic stability.

The apparent death yet continuous acceptance of neoliberalism has led many authors to examine how neoliberalism has survived and at times thrived during and after the financial crisis of 2008. Colin Crouch (2011) attributes the 'strange non-death of neoliberalism' to the dominance of the corporation in public life. Crouch (2011) argues that, unlike liberalism, competition is not the only focus and aspiration of neoliberalism but more generally the latter wishes to establish a consumer orientation and culture to all individuals and organisations. The result is the preferential treatment and subsequent domination of large commercial organisations at the expense of anti-trust laws, trade unions and social welfare.

Even though neoliberalism refuses to die, it does not mean it has remained intact from the pressures of national populism, decreasing standards of living, repetitive economic and political crises, and more recently from the pandemic outbreak. 'Libertarian paternalism' (Hansen, 2016; Gane, 2021), 'Ordonationalism' (Geva, 2021) and 'authoritarian liberalism' (Chamayou, 2021) are the most recent intellectual attempts capturing the historical trajectory and current state of politics, economy and society. These attempts and their accompanying terminology are not mutual exclusive even if they correspond to different geographical settings. Different variations of neoliberalism and what has been described as post-neoliberalism can be found in the same national political and economic settings. Post-neoliberalism does not essentially signal the end of neoliberalism but rather points towards the struggle between actors and ideas such the state, corporations, trade unions, nationalism, xenophobia and the free market to assert their dominance over the character and management of the economy, politics and society. In our book post-neoliberalism is not a term diametrically opposite to neoliberalism but rather a chronological and geographical condition capable of redefining and reassessing values, institutions and policy. The prefix 'post' indicates that such a condition is temporary after but not necessarily over the dominance of the market as the central organising principle of economy and society.

Building on these discussions of the conceptual, political and empirical evolution of neoliberalism, the book argues that the agility of neoliberalism and the very art of neoliberal governance creates its own subjectivity by resurrecting Homo Oeconomicus, but also by giving birth to its alter-ego and rival: the xeno Homo Oeconomicus. Fleming (2017) declared Homo Oeconomicus dead based on the premise that human beings are not by nature 'dollar hunting', exclusively self-interested and desperately seeking to maximise (financial) resources. We do not disagree with this premise: having or fostering a self-crippling work ethic, working excessive hours or obsessing about work is not necessarily part of human nature. But survival is and when survival is threatened by the scarcity of decent work, by the normalisation of austerity measures and by incessant politicised rhetoric that foreigners and/or machines will render you obsolete as worker, then the mythical figure of Homo Oeconomicus is reborn, or at the very least reinvigorated. The resurrected Homo Oeconomicus is not driven by greed but by anxiety and neurosis. In a society that continues to fetishise work and work ethic, and where most of us will struggle to disentangle our professions from our identities, the Homo Oeconomicus is becoming increasingly anxious about what would happen with their meagre chunk of rock. Thus, the British Homo Oeconomicus is trapped in an inescapable existential condition whereby one is constantly comparing the world as it is with the world one thinks it should be. Isin (2004: 223) refers to this as the 'neurotic citizen'.

The neurotic citizen, the argument goes, wants the impossible: absolute security, safety, tranquillity. Yet since the subject has also been promised the impossible by politicians and media alike, it cannot address its illusions and frustrations (Isin, 2004). As a result, all of the subject's wants are transformed into self-entitlement and rights: the right to secure employment, the right to tranquil living without crime and nuisance, the right to free healthcare and so on. This is where the xeno Homo Oeconomicus comes into play. The figure of the xeno Homo Oeconomicus offers a compelling explanation as to why the British Homo Oeconomicus is not getting what the latter believes s/he is entitled to; it externalises and justifies the failures and deficiencies of the British Homo Oeconomicus, as well as the absence of viable opportunities and alternatives. In order for these subjects to survive in the neoliberal market, they have to abandon their rights and liberties and to constantly compete against each other, while also trying to figure out who is stealing their jobs (Pencheva and Maronitis, 2018).

The structure of the book

The book coheres around a series of case studies, theoretical and policy debates, drawing on a wealth of academic and non-academic literature. Automation and immigration appear in this book, both as an objective medium through which British economy and society can be observed and analysed and as conceptual tools for the purpose of challenging political narratives, and ideological perceptions of individuality, freedom, equality and community.

Chapter 2 examines how the competitive labour market constructs a specific reality in which a political and social subject is formulated, developed and transformed. It is within this intersection between the promise of clamping down on low-skilled migration and the structural need for migrant labour that we focus our attention on Homo Oeconomicus and reconceptualise this subjectivity along the lines of race and ethnicity as British Homo Oeconomicus and xeno Homo Oeconomicus. Contrary to recent proclamations of the death of Homo Oeconomicus and of assertions over the power of neoliberalism to eradicate racial discriminations, we focus on the symbiotic relationship between neoliberalism and racism and the former's ability to adapt and thrive in social and political conditions which might appear as incompatible with its intellectual roots and political and economic objectives.

Chapter 3 deploys the concept of resilience in order to analyse the role inequality plays in the competitive labour market. Focusing on the nationwide campaign Pick for Britain that aimed at recruiting domestic horticultural and farm workers and on the Thank you Amazon Teams campaign praising drivers and warehouse workers, we argue that the good of the nation and the good of the corporation converge in the interpellation of the worker as a subject that needs to respond and adapt to demands for lower immigration and higher productivity while accepting inequality as an inevitability of an economy incapable of reconciling immigration and automation with decent working conditions.

Chapter 4 tells the story of how technological development and automation of work dominated political thinking and policy. It tells the story of technological fear, suspicion and inevitability. The chapter overviews and examines a wide range of policy documents and reports on automation, robotics and technological displacement with a theoretical framework provided by Marx, Polanyi and the *operaismo* movement. We argue that the competitive relationship between robot workers and human workers framed by the principles of labour cost, efficiency and productivity results in the shift from the integration of workers as a collective in a volatile social and economic environment to a project of self-realisation by establishing links between performance, knowledge and the ability to remain employable in a competitive automated economy.

Chapter 5 adopts a genealogical approach in exploring questions about the very purpose of immigration policy and whether it is possible that a government can be truly pro-immigration. The chapter explores British political and policy developments spanning from the early 20th century to the current post-Brexit context. In so doing, it outlines three main discourses that have come to define British migration policy: racialisation

(1900s–1980s), technocratic pragmatism (mid-1990s–2010) and the current securitised discourse that facilitated Brexit and which continues to drive post-Brexit politics and policy. Thus, the chapter develops the argument that migrant workers could be understood as robots in the sense that they tend to be defined by their economic value and utility while also being denied their humanity and social needs. The argument is supported by a wide range of political, policy and empirical evidence, that is, key immigration acts and policies, Migration Advisory Committee (MAC) reports, as well as social policy changes under the Coalition and Conservative governments, which aimed at maximising the economic gains of migration while minimising social expenditure.

Chapter 6 argues that employers in sectors with a high concentration of migrant workers are most likely to continue to rely on such precarious migrant labour, despite pre- and post-Brexit promises for increased investment in automation in labour-intense and migrant-dominated sectors of the economy, such as agriculture. Empirically, the argument is supported by examining specialist reports, political and media statements, as well as the Pick for Britain campaign as a case in point because it exemplifies a politically salient friction between the long-standing racialisation of EU migrants and dependency on their labour. By critically engaging with the trope of cheap labour we show how it co-exists within a discursive reality where the insufficient deployment of automation technology in the agricultural sector clashes with the significant reliance on precarious and exploitative migrant labour, which is progressively dehumanised by post-Brexit migration policies.

Chapter 7 serves a dual purpose. It offers a reflective summary of our main analytical findings, while also offering a new level of analytical abstraction by examining the political and theoretical alternatives to neoliberalism. In this chapter we critically examine and contrast the nostalgic approach of state interventionism and of homogenous traditional communities which became powerful campaign devices for the right-wing parties with the futuristic approach of UBI, full unemployment and the constitution of communities independent of the structures and habits of work. By accepting the optimism of a future where work ceases to be a constitutive component for both individual and collective identity, we argue for a new configuration of the common that is able to be critical of itself through constant renewal and rejection of a social political order based on race and skills.

2

The Re-birth of Homo Oeconomicus: Self and Other, Immigrants and Robots

'Rethink, reskill, reboot'

Over the summer of 2020, soon after the easing of the national lockdown restrictions, the topic of work and more specifically of how we work, where we work, with whom we work, and how important work is for our status, creativity and mental health became inseparable from the discourse on the pandemic and public health. Media across the ideological spectrum in conjunction with a significant number of MPs and other political figures urged British people to return to their offices and other work-related spaces. A series of events as well as public anxieties informed these official and unofficial campaigns to return to work (*The Economist*, 2020a; Tett, 2020): the office exodus and the devaluation of real estate in cities; the disappearance of the 'human touch' and 'human interaction' in business decisions and actions; the negative effect on businesses serving office spaces and office employees such as retail, taxis, hotels and catering; and of course the rather persistent existential questions over the meaning and value of work: How can workers be supervised if and when they work from home? Are workers more or less productive when they don't have to endure long and unnecessary commutes? If certain jobs can be done from home, does this mean they can be outsourced abroad to workers with lower salaries?

Keeping up with the emerging COVID-19 discourse, it is important to notice that these anxieties apply to a specific set of workers who have been classified by the government as 'non-essential' workers. In other words, they apply to those whose work is not essential to the fighting of the pandemic such as traders, lawyers, financers, university lecturers, artists and consultants of any kind. On the other hand, workers whose physical presence is required in the workplace such as carers, nurses and supermarket staff were classified as 'essential' workers. In a way, that was a belated act of recognition of people receiving the minimum wage and working under conditions of constant uncertainty generated by hostile immigration policies, underfunding of services and failed privatisation schemes. Prior to the government's response to the pandemic, this specific group of workers found themselves in the contentious and problematic category of low-paid workers whose salary does not meet the £35,800 minimum salary threshold for migrants to settle in the UK.

The vote to leave the EU in 2016 and the landslide victory of the Conservatives in the 2019 general election with the slogan 'Get Brexit Done' not only associated immigration with the country's fragile cultural cohesion but also with the decreasing incentive of British employers to train British workers instead of relying on low-paid immigrant workers. Once the new category of 'essential workers' emerged the government reduced the salary threshold by 30 per cent. Subsequently, immigrants on salaries of £20,480 but with enough points under a so-called Australian-style immigration policy will be able to qualify for jobs where there is shortage of workers and will be entitled to get UK citizenship after six years of continuous residence. Besides the changes in public attitudes and immigration policy, the newly found social and economic category of the 'essential worker' indicates a change in the definition and understanding of work during the pandemic. The low-paid work of the essential worker has been portrayed by local and national governments as quintessential 'communal work' (Komlosy, 2018) that consists of tasks and duties workers have to perform in order to help the community as a whole during a moment of public health and economic crisis.

The management of this public health and subsequently economic crisis required a national consensus between the working class and the middle class, between the essential and non-essential workers. This consensus relied on the classification of essential workers as heroic and altruistic and on the capacity of technological means to facilitate remote working and education and implementation of law and order. Put differently, the management of the pandemic relied on the sung heroes, often immigrants, in the newly categorised essential sectors of the economy and on the unsung heroes of the established yet problematic and controversial automated economy. Boosted by corporate and governmental responses to the pandemic, the latter category signalled in the most boisterous manner that the future is here, and existing technology can actually deliver the utopian or dystopian reality of an automated economy. Even though the experience of deploying robots in disasters and crises might vary, the advent of robots in the pandemic was not framed as a hostile takeover of jobs but as a collaborative relationship – a collegial helping hand performing tasks humans are not able to do or that they need assistance with. Put differently, the pandemic and the subsequent governmental and corporate responses to it reframed the discourse on automation and eased, at least temporarily, the fears of a massive loss of jobs.

The latest frontier of automation and robotics has been the service sector that predominantly occupies the highest percentage of workers in high-income countries. For years, the gradual disappearance of manufacturing jobs to automation and robots has been perceived as an inevitability and a technological solution to a human and economic problem – that of efficiency and productivity. Technological evangelists and automation advocates often present automation of manufacturing as a tried and tested process that could and should be implanted in the service and knowledge economy. Yet, the apparent social protection of the middle class from the technological and economic changes is of paramount importance for maintaining a sense of a consensus over the management of the pandemic. In economies largely defined by knowledge and services, the middle class is indistinguishable from political understandings of prosperity, equality and meritocracy. Even though policymakers and political actors in general systematically avoid defining the middle class in terms of income, occupation and lifestyles, they increasingly view it as the dominant social group through which the whole of society could and should be viewed.

As soon as the size of the economic downturn and the rising unemployment figures became available the fragile consensus of the pandemic had to be redefined. The Chancellor of the Exchequer made it clear that not all jobs can and will be protected but the government has the duty to provide the optimal conditions for people to retrain and reskill. The new interfaces for remote working such as Zoom and Microsoft Teams together with the use of robots in health, retail, and law and order have rekindled the government's interest in coding and software engineering and other related skills. Indicative of this persistent fascination with and interest in the digital economy is a government-backed campaign that encouraged people who work in the arts to retrain for a job in cybersecurity (Jordan, 2020). The campaign depicts a young woman named Fatima in ballet-wear, fastening her ballet pumps with the caption 'Fatima's next job could be in cyber. (She just doesn't know it yet)'. Notwithstanding a barrage of jokes and numerous parodies of the campaign concerning politicians and their skills and future careers, the campaign's intention was to declare the government's welcoming attitude to cybersecurity and its prioritisation over the arts and the general culture sector. Even though the campaign was characterised as 'crass' and leading government officials distanced themselves from both the content and intentions of the campaign, we cannot possibly ignore the manner in which the campaign encapsulated both the spirit and the objectives of neoliberalism, and more specifically of the competitive labour market. Fatima, the protagonist of the government's failed campaign, is addressed as an individual competing with other individuals for either a place in ballet or in cybersecurity. Fatima's next job could be in cybersecurity, but she just doesn't know it yet. Such a job is presented in the campaign as a new, exciting and most importantly as an irreversible, unstoppable force towards the future of employment. Further to the prioritisation of cybersecurity over the arts epitomised by the slogan 'cyber first', the campaign praises the merits of individual action and responsibility: 'rethink', 'reskill', 'reboot'. Despite the fact that this a government-backed campaign, it is Fatima's responsibility to rethink her career, develop new skills and consequently transform her life by demanding a higher salary in a competitive employment environment. The government is here to encourage the employed and the unemployed or to give them a nudge but by no means to support a career in ballet.

Should Fatima be worried? Should she start feeling insecure about pursuing a career in ballet? Should Fatima be aware of the competition in either ballet or cybersecurity? Should Fatima be thinking of funding her training for a career in cybersecurity? The anxiety about work prospects, the insecurity concerning the viability and applicability of skills and knowledge are not only reflective of a political and economic system sustained by competition and by addressing the employed and the unemployed as isolated individuals, but also of the capacity of that system to construct a subject that is rational yet helpless, heroic yet insecure.

The subject in/of the competitive labour market

It is the intention of our book to examine the current manifestations of neoliberalism and of neoliberal rationality through the prism of the competitive labour market – a space where jobs are won and get stolen, a space where regulation and deregulation are at once vital for its survival. Neoliberalism and more specifically the liberalised competitive labour market have always been associated with destructive tendencies and qualities. Curbing workers' rights, limiting the role and power of trade unions, lowering wages for maximising employment figures are all constitutive parts of what is known as the competitive labour market. However, we cannot grasp the political and social complexities of the competitive labour market if we limit ourselves to neoliberalism's destructive qualities and ignore the productive ones. The emotional states of fear, loss and anxiety produced by the discourses of automation and immigration do not necessarily point to a political and economic system in disarray. As a matter of fact, financial crises and market failures, poverty and political corruption indicate that neoliberalism's success does not exclusively depend on economic outcomes but on establishing and presenting a political and economic logic of competition as the only doctrine for achieving prosperity and growth. We have to contextualise fear, loss and anxiety as part of a specific motivational strategy and a wider governmental technology. Michel Foucault (2004) in his Collège de France lectures argues that neoliberalism is an all-encompassing system in which new subjectivities emerge. The way these subjectivities conduct themselves and interact with each other in a competitive environment reveals the political aspirations and limitations of neoliberalism. For Foucault (2004) neoliberalism is a political rationality that aspires to produce a permanent consensus among all those who operate in its confines. Industrialists, financiers, manual workers, service employees, business executives, politicians, policymakers and law and order enforcement have to operate according to the principles of the neoliberal rationality. This consensus requires and at the same time manifests itself with the construction of a collective subject that transcends class, gender and race without necessarily aiming to dismantle them as social categories. The point here is not to observe if and how different people can be grouped together by the opportunities, motivation and managerial techniques of the labour market but instead to analyse how the market constructs a specific reality in which a political and social subject is formulated, developed and transformed.

Consequently, for Foucault (2004) and later for Dardot and Laval (2013), the market constructs its own subject where individuals have to learn a code of conduct that supports and is being supported by the market. This subject is historically known as Homo Oeconomicus and has been a permanent feature of political and economic thought. If Homo Oeconomicus is constructed and understood by the discursive and institutional practices of neoliberalism and the competitive labour market, then it would be reasonable to assume that its analysis will be informed by the analytic qualities of structuralism. However, at this point we will have to take into consideration the theoretical accounts of the proponents of competition and the free market whose depictions of Homo Oeconomicus paint a completely different picture to the one provided by Foucault, and Dardot and Laval. Following the theoretical elaborations of Michel Wieviorka (2012) and Alain Touraine (2000, 2010), we can argue that the depiction of Homo Oeconomicus by the theoretical proponents of free markets and competition is closer to an achievement and a product of constant struggle against institutions, which seek to regulate the market and strike a balance between individual and collective interests. The self is attached to or rather becomes a project of entrepreneurial activity and individual freedom.

For Wendy Brown, such a project is informed and conditioned by two parallel replacements involving labour and productivity. First, as soon as competition becomes the market's guiding principle, all market participants regardless of their respective professional qualifications and capacity, for example consumers, producers and workers, are considered to be human capital. As part of the market's human capital, every participant is and has to be an entrepreneurial subject competing with other entrepreneurial subjects. The mass transformation of labour to human capital and the subsequent categorisation of market participants as competitive entrepreneurs challenges existing notions of class structure and inequality. Parallel to the replacement of labour by human capital and the transformation of market participants to entrepreneurial subjects, the notion of productivity has given its place to entrepreneurship. According to Brown (2015), this replacement indicates that the enterprise society envisioned by neoliberal thinkers is not necessarily a society structured around the production, consumption and exchange of commodities but rather around competition. In effect, the neoliberal subject Homo Oeconomicus is the subject of enterprise whose objective is self-investment.

Yet, self-investment does not necessarily mean that Homo Oeconomicus as an entrepreneurial subject is free from the supervisory powers of the market and the control of the state. Self-investment does indeed signal a break with the Keynesian tradition of state investment and central planning but at the same time it creates new structures of control and discipline that derive from and address the individual.

How can self-investment be evaluated? Without social protection and secure employment Homo Oeconomicus has to acquire and by extension demonstrate its employability skills and creditworthiness. Both employability skills and creditworthiness become an efficient if not inaccurate means to assess human capital according to the Homo Oeconomicus' capacity to perform multiple tasks and generate income though work and credit. The presence of Homo Oeconomicus and its subsequent assessment signal the individualisation of work and wage relationships. For Bourdieu (1998), the ubiquity of individual career paths, performance evaluations, salary increases and granting of bonuses are symptomatic of a particular type of labour contract that ensures at once the interpellation of workers as independent from the aegis of firms and their self-exploitation as solely responsible for the survival and growth of such firms.

Kluge and Negt (2014) keep reminding us that Marx demonstrated in *Capital* that social wealth is inextricably linked to the impoverishment of the individual worker. For Marx and by extension for Kluge and Negt (2014) individual workers selling their labour and investing their skills and time in the workplace do not have a stake in the wealth they produce. Instead, what they do get in return is the 'minimum self-representation' they need in order to sustain and reproduce their existing living conditions. Kluge and Negt (2014) are quick to note that impoverishment is not limited to physical and bodily manifestations such as hunger but more accurately designates a specific kind of social relationship where the possibilities for change individuals have at their disposal are minimal or even non-existent. While Marx's analysis in *Capital* focuses predominantly on 'dead labour', in other words on machinery, policies and socioeconomic relations, Kluge and Negt (2014) shift the focus on 'living labour'; that means on workers as the actual subject of capitalism. Following Raymond Williams' (2020: 40) dictum that 'the most important thing a worker ever produces is himself … the broader historical emphasis of men producing themselves and their history' we deploy Homo Oeconomicus as the subjectivity and embodiment of neoliberalism. Can neoliberalism say 'I' and if so, how does it say it?

Theorists of neoliberalism and of the entrepreneurial subject do not necessarily deviate from the Marxist diagnosis concerning the economic and social status of the individual worker. However, the solution offered points towards further fragmentation, or to be more precise individualisation of the workforce, accompanied by the much-desired weakening of social structures. In effect, Homo Oeconomicus can only thrive as long as they realise they have to exist in opposition to regulations, institutions and central planning.

The central intellectual figures of the struggle against institutions, central planning and regulation are Joseph Schumpeter (2010) and Ludwig von Mises (1998). While Homo Oeconomicus provides a common platform for developing a narrative in favour of free markets and against regulation, there are noticeable differences between the two. Most notably, Schumpeter (2010) foresaw the death of entrepreneurial man and subsequently the end of capitalism with the rise of corporations, bureaucracy and managerial culture. At the heart of his argument lies the need for optimal political and economic conditions for the entrepreneurial man to thrive. Schumpeter (2010) became suspicious of the direction of economic progress and in particular of the corporate structures and managerial culture that advocated coordination, cooperation and planning. Paradoxically, Schumpeter (2010) believed that the ascendancy of global corporations will restrict the role of the entrepreneur and eventually cause the end of capitalism. According to Schumpeter (2010), regulations, complex policies, corporate structures and managerial culture are a hindrance to individual entrepreneurs whose skills and innovations are the genuine contributors to economic growth and progress. Schumpeter's conception of the heroic and risk-taking Homo Oeconomicus is economically and socially shaped by the early entrepreneurs of the Industrial Revolution. Determination, ingenuity and audacity are some of the qualities required for Homo Oeconomicus to thrive in an environment dominated by competition and uncertainty. However, such qualities, Schumpeter argues, are not necessarily available to all people but belong to few exceptional individuals whose actions and economic presence classify them as leaders. In Schumpeter's (2021) *The Theory of Economic Development*, Homo Oeconomicus neither seeks to build a safe economic environment nor aspires to a socioeconomic consensus among relevant economic actors and market participants. On the contrary, the Schumpeterian daring individual prioritises rupture over consensus and mobility over stability. Unlike most neoliberal thinkers of the Austrian theoretical economic tradition, Schumpeter expressed a disdain for the state as the guarantor of stable legal framework in which economic activity can take place.

For Mises, on the other hand, these structural changes in corporate governance and culture are an integral part of the complex landscape Homo Oeconomicus has to operate in. Mises did not share Schumpeter's pessimistic predictions but instead focused on the performative qualities of Homo Oeconomicus. For Mises, anyone can be Homo Oeconomicus – an entrepreneurial man operating in complex and often hostile environments in search for profit. The success of Mises' Homo Oeconomicus performance can be found in the management of these conditions and of course in the maximisation of his/her profits. For Mises the formulation of a middle way between entrepreneurship and state intervention is politically and socially hazardous, mainly due to the detrimental effects state control has on the actions and choices of the entrepreneurial citizen. State intervention, according to Mises, creates dependent beings lacking imagination and most importantly entrepreneurial skills. Individuals are perfectly capable of pursuing their interests without the help of the state, and all economic science can be stripped down to human action – a 'praxaeology' in Mises' lexicon.

The relationship between Homo Oeconomicus and the environment they have to operate was formalised by F.A. Hayek (2006a) as a relationship between freedom and unfreedom. While Hayek finds himself in agreement with Schumpeter and Mises that Homo Oeconomicus has to overcome the objective difficulties presented by policies, governmental planning and social attitudes, the dangers of socialism cannot be ignored, mainly because it will take away freedom and eliminate the entrepreneurial spirit. Socialism for Hayek is dangerous not necessarily because of its aims, namely social justice, security and greater equality, but because of its means to these aims (Gray, 2013). For Hayek the social method for achieving these aims results in the abolition of private property, enterprise and 'the creation of a system of planned economy', in which Homo Oeconomicus is replaced by a central planning body (Hayek, 2006b: 34). The objective for Hayek (2006b) is to protect the 'price system' from state interference because it enables entrepreneurs to freely observe the fluctuation of prices and act accordingly in order to maximise their profits. It is worth noting that for Hayek (2006b) the actions of Homo Oeconomicus are the actions of a well-informed, objective and sober individual similar to the engineer who watches the hands of a few dials. For the price system to function properly and for competition to prevail, Homo Oeconomicus must be able to adapt and at no stage be able to control the system. Hayek insists on the existence of complexity and fragmented knowledge for the survival of the price system: 'the more complicated the whole, the more dependent we become on the division of knowledge between individuals whose separate efforts are co-ordinated by the impersonal mechanism for transmitting the relevant information known by as the price system' (Hayek, 2006b: 52).

Central to Hayek's (1973) theoretical plan of the function of the state in a free-market economy is his formulation of the rule of law. Considering Hayek's aversion to central planning and socialism, his argument for a limited state is structured around a creative ambiguity over the latter's role and reach. Instead of opposing the state as matter of liberal principle, Hayek re-engages with the classic liberal debate on the state and more specifically with the external and internal mechanisms for its regulation. While acknowledging the state as the legitimate protector of all citizens, its reach must not compromise individual freedom and economic activity. Inevitably, Hayek's (1973) attention turns to the welfare state as the source of the collective misunderstanding about freedom and protection and of all ills in the 20th century. In particular, the redistributive actions of the welfare state disregard the desires and instincts of groups of peoples who will always want something different for themselves. Effectively, redistributive politics for Hayek compromise the very liberal foundations of modern societies. The Hayekian rule of law constitutes a response to what he identifies as the welfare state's unprecedented and illegitimate coercion of its citizens. For Hayek (1973), the legitimacy of the state should derive from the construction of a legal framework that remains abstract and is equally applicable to all citizens regardless of their class. In turn, Hayek (1973) insists that this is where the reach of states and governments should end; within this legal framework individuals should act as they see fit and use the resources made available by the state according to their own interests.

What are the means for by-passing central planning, avoiding the totalitarian control of the state and at the same time excluding movements opposing capitalism? Neoliberal thinkers found an answer to this question in the political theory and programme of Carl Schmitt (Kelsen and Schmitt, 2015) – a member of the Nazi party and fierce critic of liberal democratic constitutional politics (Hayek, 2006a; Chamayou, 2021). Schmitt of course could not possibly entertain the thought of dismantling the state for the establishment of a free-market economy but instead argued for the former's recalibration and qualitative transformation. Crucial to his idea and political programme was the distinction between the 'quantitative total state' and the 'qualitative total state' (2015). The latter is on a mission to concentrate in its services and agencies media technologies and the military for the suppression and, when possible, elimination of external threats and dissenting acts in its interior. In Kelsen and Schmitt's (2015) 'qualitative total state' the economy plays a crucial role for determining the domains of politics and enterprise and managing their subsequent relationship. The purpose is to depoliticise the economy by maintaining state control of key industries such as the army, transport and telecommunications and allowing everything else to be dictated by the free market. In other words, Schmitt's political programme constituted a reassurance to German industrialists and other employers that the state's authoritarian approach to eliminating dissent and ideological plurality would never extend to the domain of the economy.

Contrary to the laissez-faire economics and prevalent state phobia of Schumpeter and Mises and by taking a lead from Schmitt, Hayek (2006a, 2006b: 50, 86, 113) argues for a permanent legal framework that enables the individual to plan with a degree of confidence and reduces human uncertainty as much as possible. Hayek's references to the law and human uncertainty indicate that neoliberalism is not a destructive force threatening institutions and the social fabric for the establishment of the rule of the market. Most importantly, Hayek's argument points to the 'denaturalisation' of markets and competition (Foucault, 2004; Gertenbach, 2017). Competition in national and global markets, according to Hayek, should not be perceived as some form of primordial act that is being restrained and distorted by state regulation but rather as a political artifice whose function is to create a specific code of conduct in line with the benchmarks of economic growth and profit maximisation.

The emphasis on competition by Schumpeter (2010), Mises (1998) and Hayek (2006a, 2006b) suggests a shift in the way Homo Oeconomicus operates as well as is perceived. In liberalism, Foucault points out, Homo Oeconomicus was the 'partner of exchange and the theory of utility based on a problematic of needs' (2004: 225). According to this conception, the market is a social space where participants offer what they have, in exchange for what they need. In the neoliberal logic of Schumpeter, Mises and Hayek, the market as the place of exchange is transformed into a place of competition where the participants are not necessarily interested in exchange but in 'investing' in themselves as both producers and consumers.

The emergence of Homo Oeconomicus as the defining subject of neoliberalism comes at great cost to the political action of human beings and their sense of security. The economic value of individuals and the training required as self-investment in order to thrive or even survive in the competitive labour market eliminate their political dimension and ability to actively participate in a democratic polity. In other words, Homo Oeconomicus establishes a clear hierarchy concerning the culture and actions of human beings – a hierarchy where economic activity not only sits at the top, but most importantly informs and determines all other activities. However, it would be naive to assume that such an activity remains unsupervised and unclassified. As Michel Feher (2021) argues, there exists a discrepancy between the promises of neoliberalism and the socioeconomic reality that it creates. For Feher (2018), the emergent 'rating culture' of neoliberalism epitomised by the presence and interference of powerful rating agencies, corporate audits and work performance reviews is at odds with the neoliberal ethos of entrepreneurialism and individualism. The contradictions between economic freedom and social subjugation, entrepreneurship and ratings, individualism and work performance reviews enable us to analyse Homo Oeconomicus as a flexible subjectivity always responding to as well as shaping other political, social and economic spheres such as immigration and automated technologies of labour. It is because of this flexibility that Wendy Brown (2015) suggests that Foucault's conceptualisation of the neoliberal subject needs to be updated in order to capture new power dynamics between the individual, employment and the market: 'Put it differently, rather than each individual pursuing his or her own interest unwittingly generating collective benefit, today, it is the project of macro-economic growth and credit enhancement to which neoliberal individuals are tethered and with which their existence as human capital must align if they are to thrive' (Brown, 2015: 84).

Neoliberalism demands from economic actors to trade off their rights and liberties for access to and participation in the market. At the same time, the state's input is not limited to the design and implementation of a legal framework that allows the expansion of the market in all spheres of life but rather extends to the minimisation, and if possible, the elimination of all opposing forces to such expansion. Is this a price worth paying? Such a question does not necessarily demand a definitive answer but more accurately the re-examination of Homo Oeconomicus and its relevance in the competitive labour market. For Peter Fleming (2017) Homo Oeconomicus is dead because it has always been part of a fictitious political and economic narrative that praises rationality, efficiency and most importantly self-interest. But, in Fleming's (2017) view, the proclamation of the death of Homo Oeconomicus is not only based on the hypocrisy of the entrepreneur but also on 'the expansive majority' of workers on average income jobs who feel powerless, frustrated and insecure. Lower salaries, meaningless tasks and the constant threat of unemployment debunk the myth of the heroic knowledgeable subject and reveal the ugly truth of a political and economic system that relies on the state's assistance for the systematic exploitation of workers and the elimination of opposing views. On a different yet similar note, Brown does not proclaim the Homo Oeconomicus dead but highlights its changing *raison d'être*: from the pursuit of self-interest to survival and sacrifice in a political and economic order that disregards notions of security and employment stability.

The new Leviathan and the management of precarisation

Is it possible to speak of a social class formed around the common experience of competition and insecurity? Is it possible to speak of a social class created in the shadow of the defunct myth of the Homo Oeconomicus? Rapid turnovers, fixed contracts and the proliferation of part-time and gig employment have become the rule to the detriment of big sections of the population. Mike Savage et al (2015) and Hugrée et al (2020) in their extensive research on class in the 21st century and Europe respectively, argue that the class most closely associated with this new social and economic normality is the working class. By introducing the precariat in their respective analyses, both sets of sociologists wish to depict the contemporary state of the working class and at the same time differentiate between a traditional working class with increasingly limited opportunities to unionise and an emerging working class comprised of people occupied in catering, care, food and parcel deliveries living under permanent uncertainty and anxiety over their living and working conditions. As a number of studies have demonstrated, those predominantly affected by the emergence and gradual establishment of the precariat are women, immigrants and young people (Murgia and Poggio, 2014; Zheng, 2018; Flores Garrido, 2020). The very existence of the precariat does not only indicate the prevalence of hostile working conditions and low payment, but also the impossibility of precarious workers being formally integrated in the labour market and receiving the social protection they are entitled to. In other words, the precariat oscillates between visibility and invisibility, formality and informality. However, it is important to stress that the experience of insecurity and the flexibilisation of work are not confined to the working class and we need to emphasise precarisation as a process that incorporates both middle- and working-class professions and aspires to normalise insecurity and flexibility across the social spectrum. Directing our attention to the Organisation for Economic Co-operation and Development's (OECD) Employment Protection Index, we see that the UK has one of the weakest employment protection policy schemes concerning individual and collective dismissals (1.23) and in fact has been getting weaker since the beginning of the 2010s. Workers in the UK are significantly less protected than their counterparts in Germany (2.50), France (2.52) and Italy (3.02). Furthermore, when it comes to the protection of temporary contracts the metrics are equally bad. In the relevant OECD league table, the UK is second from the bottom (0.38) but since the beginning of the 2000s the figures have remained the same. Even though the OECD figures and corresponding league tables paint a bleak picture for workers in the UK, US, Canada and Ireland, among others, it becomes evident how deregulation of the labour market and in effect the precarisation of workers have been deployed as means to economic growth, higher productivity and higher employment rates. Considering its scope, the process of precarisation might not be entirely successful but it certainly creates a social order – a precarity that we argue is based on racial and ethnic lines.

At this stage, it is important to underscore the critical as well as constructive role inequality plays in the neoliberal order. The competitive labour market does not aspire to the creation of a level playing field of insecurity. As Hayek

(2006a: 65) notes, 'people in general do not regard mechanical equality of this kind as desirable'. Equality, even the undesirable kind produced by neoliberalism, demotivates individuals and restrains their performance in the market. With this in mind, the experience of precarious working conditions and social insecurity needs to vary in order to encourage individuals to be more productive and discourage the formation of a social class challenging the neoliberal order. In other words, Homo Oeconomicus, even as a myth, has to be sustained.

Compliance with the political and economic system no longer depends on a strong state capable of protecting its citizens from insecurity and fear. Instead, the new art of governance involves the precarisation of individuals and the reordering of their subsequent relations and hierarchies. In the absence of state-controlled social protection, precarisation needs to be regulated in order to remain politically and economically legitimate, and most importantly its 'extent must not pass a certain threshold such that it seriously endangers the existing order' (Lorey, 2015: 2). Preacarisation of work and working conditions points to the transformation of the role of the state from Leviathan to a manager. Whereas Hobbes in *Leviathan* argues that communities can and should be oriented around the fear of greatest evil (*summum maulum*) and only a strong, undivided government can prevent civil war and anarchy, the free market requires from the state a managerial role where danger and safety, comfort and dejection have to be carefully regulated for sustaining motivation and belief in the competitive labour market.

This development in capitalism explains the regulatory character of governments and institutions. The liberalisation of the market, Polanyi argues, generated a 'countermovement' of interference by governments and institutions respectively in the form of health and safety measures, labour standards and social welfare. While this theorisation is useful for asserting the political significance of social democracy and the need to control the otherwise uncontrollable capitalist forces, it fails to acknowledge that in neoliberalism the government and the market develop a symbiotic relationship in which the former has to serve the latter by creating the appropriate environment for Homo Oeconomicus to operate.

One of the great achievements of neoliberalism and by association of the competitive labour market is that individuals, despite the uncertainty and fear they experience, hardly attribute any responsibility to capitalism itself or to themselves as active participants and supporters of the former. In effect, neoliberalism becomes an all-encompassing system that not only distorts any opposing political and economic visions but if and when possible eliminates them. Yet, one of the great theoretical and empirical contradictions is how this political and economic order is maintained. It is true that neoliberalism aspires to and up to certain extent has succeeded in creating a collective subjectivity that transcends class, gender and ethnicity for the purpose of expanding and establishing the logic of the free market in every aspect of our lives. At the same time, neoliberalism requires an *other* for demarcating the margins of capitalist societies, and justifying the trade-offs and sacrifices required for participating in the market. Traditionally, the *other* to Homo Oecomomicus could be found outside the domains of the neoliberal order and the liberalised labour market, that is the subject who lacks motivation, incentivisation and stimulation to make a success of her/his life through the structures imposed and reproduced by a dominant enterprise culture (Dardot and Laval, 2013). The new spirit of capitalism perfectly illustrated in the neo-managerial culture of risk and flexibility constructs the inflexible subject – the subject that is unwilling or incapable to perform multiple tasks, to retrain and embrace new methods of employment, monitoring and assessment as an *other* to the interest-driven, flexible and risk-prone Homo Oeconomicus. The chronically unemployed, the benefits recipient and the benefits cheat, the lazy or, less crudely put, work-shy individual, and the militant trade unionist who refuses and when possible sabotages the free market are all *others* to Homo Oeconomicus.

However, such depictions seldom refer to race and ethnicity as vital components for the constructions of an *other* to neoliberalism and Homo Oeconomicus. In order to understand Homo Oeconomicus through the prism of race and ethnicity we are deploying xeno racism as an auxiliary theoretical lens. Our justification is as much theoretical as it is empirical. On the one hand, there is an epistemological compatibility between a Foucauldian analysis of neoliberalism and Sivanandan's (2001) notion of xeno racism as a way of understanding the diverse exclusion regimes which complement the global reach or neoliberalism. Sivanandan argues that the endless drive for economic growth and profit maximisation seeks to maintain a physical and discursive displacement of all those it classifies as 'others' on the grounds of their work ethic and their intention to prey on the wealth of the West, harm its identity and standards of living.

But before we proceed with the social and political contextualisation and function of xeno Homo Oeconomicous we need to step back and discuss the problematic yet interdependent relationship between neoliberalism, racism and nationalism. Theoretically, the relationship between neoliberalism and racism is not only odd but outright contradictory. How can a political and economic theory advocating freedom, deregulation, disentanglement from state structures and centralised planning be compatible with the policing and exclusion of people based on their ethnicity, race and nationality? It was Milton Friedman (2002), after all, who considered the competitive market as the creator and guarantor of a post-racial economic environment that encourages its participants to leave behind their established cultural rituals and dispositions for the purposes of greater flexibility and efficiency. For Friedman (2002), one of the most positive effects of the free market is the protection that it offers to its participants from discrimination against their economic activities for reasons that are unrelated to their productivity. In a competitive free market, Friedman (2002) insists that discrimination on the basis of race and gender is diametrically opposed to the corporation's drive for profit maximisation. In other words, corporations can either choose to exclude people or maximise their profits. This argument does not only contribute to the assumed reciprocal relationship between capitalism and freedom but more importantly makes the case for the moral character of the free market. Friedman's (2002) conception of the free market encourages competition and at the same protects vulnerable groups from discrimination and exclusion.

Based on Friedman's view on the assumed antithetical relationship between capitalism and discrimination, Gary Becker (1957) placed exclusion and discrimination at the core of economic theory and policy. Becker's theory is driven by and targeted at the persistent social issue of the wage differential between different racial groups. By avoiding normative claims regarding social justice and ethics, Becker advances the argument that Homo Oeconomicus is quintessentially a rational and social subject. The market which constitutes Homo Oeconomicus' natural terrain is where people pursue their own interests and interact with each other. Becker's economic argument as well as theoretical formulation rely on the economic notion of equilibrium – a point at which the market participants must balance their own interests with their mutual social interaction. Inevitably, for Becker discrimination affects both the discriminators and the discriminated, albeit in different ways. If racial discrimination imposes lower wages for Black workers in relation to similarly qualified White workers, then the discriminator-employer will have to pay more to retain an exclusively White human workforce. For Becker (1957) this creates two costs: first, discrimination is expensive because discriminatory employers will have to pay more in order to retain a productive workforce while excluding others in line with their discriminating beliefs. Second, Becker (1957) predicted that the competitive labour market will eventually resolve the social problem of discrimination by pushing discriminated workers out of hostile work environments and towards hospitable ones. In effect, Becker managed to frame the charged social issue of discrimination and exclusion as an economic problem of supply and demand with the labour market as the ultimate problem solver.

The current political and economic conjuncture not only renders Friedman's and Becker's accounts of neoliberalism's effect on culture as over-optimistic but more importantly it illustrates the symbiotic relationship between neoliberalism and racism and the former's ability to adapt and thrive in social and political conditions which might appear as incompatible with its intellectual roots and political and economic objectives. The traditional targets of neoliberalism such as welfare dependency and state planning are increasingly viewed through a racialised prism. The management of immigration by the EU and national governments in conjunction with the demand for cheap migrant labour have generated a series of racialised categories based on occupation, income, residence, electoral behaviour, culture and values. Recent manifestations of racism are entangled with these categories but more importantly with the 'dethronement' (Brown, 2019) of a specific type of Homo Oeconomicus: the White male worker whose life has been defined by irreversible cultural trends and technological advancements. Indicative of this dethronement are the campaign slogans of populist parties and campaigns aiming at reversing the existing cultural order but not necessarily the existing political and economic order. From Marine Le Pen's 'France for the French', Matteo Salvini's 'Italians First' and Alternative für Deutschland's 'Our Country, Our Home, Our Germany' to Donald Trump's 'Make America Great Again' and the Leave campaign's 'Take Back Control', there is a demand for returning to a glorious past driven by a sense of national entitlement and cultural homogeneity.

There are three main characteristics that bind the emergence of national populism and its subsequent demands. One of the primary concerns of national populist parties and campaigns is to provide a sense of social protection by restoring a sense of pride in work and by arguing for the repatriation of key industries which have been outsourced to low-wage economies. Second, the most common target for national populists is the migrant worker, mainly because the latter symbolises both the loss of a culturally homogenised society and an unequal employment landscape in which low wages and deregulation are more preferrable than tradition, community and national culture. Finally, neoliberal economic policy is an indispensable feature of the contemporary national populist worldview. Once again there is the expectation of the state to facilitate an environment where corporations can thrive. Consequently, national success relies on corporate success and citizenship becomes synonymous with flexibility, efficiency and productivity. By situating the nation as an active competitive participant in the global economy, neoliberal nationalism operationalises the state as a complex economic unit that demands from its citizens to be productive, efficient and when necessary to sacrifice their freedom and wellbeing for the greater common good. Yet, neoliberal nationalism, like all manifestations of nationalism, cannot proceed without the basic distinctions of 'us' and 'them', national and foreigner. A major administrative and ideological problem caused by the prevalence of neoliberal nationalism is how to deal with efficient and productive foreigners who contribute to the nation's competitive edge as well as to growth and productivity targets.

The symbiotic relationship between neoliberalism and exclusionary politics indicate the desire of populist movements and parties to reinstate some kind of political and economic order based on specific national and cultural values and criteria. Neoliberalism's deregulation, uncertainty and precarity are perceived by all actors involved as unsustainable and for this reason populist movements seek to recreate an order at the limits of the free and competitive labour market. Here, we can detect the urgent need to redraw the contract between the state and the people and to redefine the latter in the neoliberal economy. The reluctance of governments and parties to challenge the current capitalist structures require a reconsideration of Leviathan and its role as an enabler of competition and exclusion. Leviathan might have turned into a manager but does not mean it cannot be a racist and punitive manager. Building on Pierre Bourdieu's anthropomorphising of the neoliberal state, Loïc Wacquant (2009) distinguishes between the Left and Right hand of Leviathan. The Left hand is activated by ministries and government agencies ensuring limited housing, social protection, welfare and public education all in line with the principles and objectives of the neoliberal economy. The Right hand is animated by and is responsible for enforcing these principles by implementing austerity programmes, facilitating partnerships with the private sector, and advocating economic deregulation. Wacquant provides a theoretical and practical extension of the state's Right hand by incorporating incarceration and penal policies. This depiction of a rather longer and stronger Right hand highlights the pivotal role of law and order and explains the mission of the state in the neoliberal economy. More specifically, Wacquant's formulation assists our understanding of a state that is more concerned with the management of precarity and the penalisation of poverty than with their eradication.

For this approach to gain legitimacy, there needs to be a distinction among national, ethnic and racial lines – between the national Homo Oeconomicus and the xeno Homo Oeconomicus. The theorisation of capitalism and its analysis in social, political and cultural contexts have strengthened the view that economic theories have the power to determine national and international policies and more importantly to organise everyday life. Yet, economic theories with aspirations of shaping relations among nation-states, international organisations and intranational polities cannot be implemented without adopting specific national and local characteristics. As Cornel Ban (2016) illustrates in his analysis of the implementation of neoliberalism in Spain and Romania, economic ideas are always open to interpretation and adaptation. Therefore, neoliberalism must always be treated as an ongoing economic and political process that alters and is being altered by local, regional and national cultures and economies. The national context in which *Robots and Immigrants* is placed demands a different analytic direction and approach. Traditionally, neoliberalism is articulated and analysed as a global rationality that is imported or even forced upon a national economy resulting in a variety of economic growth and employment targets, protests, social and political restructurings. In *Robots and Immigrants*, our starting position is that after 40 years of free-market policies, economic deregulation and demands for workers' flexibility, neoliberalism is one of the most – if not *the* most – defining features of British economy and society. Therefore, we treat any attempts to control the rationality of the free market as something culturally and administratively external to the present character and structures of the British economy and society and such attempts assume different and variant characteristics.

The vote to leave the EU and the popular rejection of liberal democracy prompted consecutive Conservative governments in Britain to amplify the relationship between state and citizen and subsequently the statutory protection against external threats. By acknowledging the precarious position of British Homo Oeconomicus the very concept of sovereignty is being redefined. Any idea of sovereignty implicitly or explicitly relies on a process of precarisation of citizens, which simultaneously reinforces the demand for strong physical, social and cultural borders and exposes the vulnerability of the sovereign subject. The anxiety experienced by the British Homo Oeconomicus and the demands for sovereignty and control do not necessarily challenge the process of precarisation initiated and propagated by neoliberalism but insist on maintaining a racial and ethnic order in which the sovereign subject is prioritised.

An economic system of precarisation legitimised by national, racial and ethnic hierarchies is not something new. As early as 1935, W.E.B. Du Bois analysed capitalism, racism and exploitation by formulating the concept of metaphorical payments and more specifically of 'psychological wages'. Whiteness, famously argued by Du Bois, became a valuable means to raise the status of the exploited White worker and establish the belief that income alone cannot measure the rewards of capitalism. Further to the concept of the 'psychological wage', Du Bois detects a deep-rooted, unquestioned belief that the world is and should remain in the property of White people. In the competitive free labour market, Whiteness is not as important as it used to be in 1930s America. The distinction between the national Homo Oeconomicus and the xeno Homo Oeconomicus is not limited to the apparent threat of asylum seekers and other groups of non-European migrants, but also extends to the assorted dangers posed by phenotypically White EU migrants. As soon as the *other* is and can be White, the distinction between national and xeno Homo Oeconomicus is to be found in the apparent reasonable expectation of priority concerning job security, better payment and access to welfare and of course in a clear-cut hierarchy concerning the ownership of jobs in a volatile and fragmented employment environment.

Raising the question in the book's title once more: who steals jobs? The neoliberal government might not be able to give a definitive answer but can confidently claim that nationality and ethnicity should play a vital role in managing precarisation in order to maintain the existing social and political order.

3

'A Necessary Evil': Progress through Normalising Inequalities and Competition

Inequality and participation in the market economy

The vote to leave the EU in 2016 and the landslide election of the Conservatives in 2019 brought to the foreground of British politics the issue of inequality and in particular regional inequality structured by the relationship between cities and towns and between economic centres and peripheries. According to the Organisation for Economic Co-operation and Development's (OECD) report on regions and cities, the UK is the sixth highest among 30 developed and industrialised countries in terms of regional economic disparities, and recorded the fourth largest increase in these disparities between 2000 and 2016 (OECD, 2020). While the report coheres around a variety of metrics and issues, the more germane to this book are productivity and employment. Not only does Greater London feature as the productivity and employment 'frontier' in the UK, but the disparities between Greater London and South West England and the West Midlands are getting wider.

Even though debates on the cultural, political and economic significance of Britain's North and the Midlands have been going on for decades, the latest disparity figures were and still are used by politicians and analysts for explaining public attitudes towards immigration and national identity, and by association the voting behaviour of a significant segment of the population previously overlooked by policies and political manifestos. The explanatory potency of these regional disparities is supplemented by the formulation of a collective subject that encapsulates cultural and material anxieties over the meaning of work and its significance in understanding national history and identity as well as a sustained criticism of technocratic and liberal politics. A series of political analysts and commentators have attributed the latest electoral results and upheavals to the dominance of managerial politics and to the limits of liberal democracy to represent the anxieties and defend the rights of people defined interchangeably as the 'left behind', 'traditional working class' and 'blue collar workers'.

Often the left behind are geographically and culturally defined in relation to their proximity to London and in particular Westminster politics. But it is now political parties, policymakers and the whole system of representative democracy that seem to be out of sync with changing social attitudes and electoral preferences. For Eatwell and Goodwin (2019) the 'backlash against liberal democracy' and the rise of 'national populism' are framed by distrust of the political elites, destruction of the nation's identity by immigration and globalisation, deprivation as result of unequal taxation and uneven distribution of wealth, and finally by the dealignment of traditional political affiliations and attachments due to the inability of parties and unions to represent the anxieties and interests of the electorate.

By focusing on the relationship between geographical exclusion and social disparities, Christophe Guilluy (2019) attributes the rise of populist politics and the rejection of liberalism to the distortion of social reality by the rhetoric of 'openness' and 'peaceful inclusivity'. In particular, Guilluy (2019) identifies two social groups in this reconfigured political and cultural environment. The first group consists of the 'new moderns', capable of navigating their lives in the complex globalised network of culture, knowledge and commerce. Convinced of the validity of their own views, the 'new moderns' are willing to teach their fellow citizens to see and experience the world as they do. The second group consists of the 'new ancients', incapable of navigating their lives in the complex global network, and marginalised by recent developments in economy, society and culture. According to Guilluy (2019), those two groups encapsulate the current contradictions as well as battles between the winners and losers of globalisation – between the new upper and lower social classes. These new dividing lines do not only position the upper class as the morally superior class but also dismiss any critiques of and anxieties over immigration, deindustrialisation and the erosion of local and national traditions as phobic and populist.

Even though both sets of theorists present their theoretical and empirical findings as a by-product of contemporary tensions between core and periphery, elites and the people, it is important to note that such distinctions and their subsequent management are as old as politics. In particular, there is a strong tradition in political thought that aspires to find an equilibrium between the elites and the people without necessarily dismantling old and existing hierarchies and discrepancies. Most notably, Machiavelli (2007) saw in the conflict between wool workers and the ruling class in Florence, known as the 'tumult' of the Ciompi in 1378, a certain kind of social and political dynamism. Despite the fact the uprising only lasted for six weeks, the fear of another insurrection and rule by artisans and manual workers informed the direction of the ruling elite for years to come. Machiavelli (2020) not only considered the causes and motivations of the uprising by examining the working and living conditions of the workers but, more importantly, studied the way in which elites had to adjust their rule in order to avoid future upheavals.

On the other hand, for someone like Naudé (2020), society's divisions were mostly intellectual and not political. Naudé (2020) identified two social groups generally defined and shaped by their respective attitudes to ideology and superstition. There were the enlightened elites, of which he was a prominent member, able to free themselves from the restraining powers of ideology and religious superstition, and there were the uneducated masses acting on instinct and unable to reason. Effective statecraft required the control of peasants and the masses by maintaining a sociopolitical environment that prevents uprisings through a system of concessions and recognition of existing hierarchies.

These theoretical and empirical observations indicate that the condition of being 'left behind' under the current structures of capitalism is a by-product of exclusion from the market economy rather than a consequence of participation in it. Against this backdrop of new class configurations, the UK Conservative government under the aegis of Prime Minister Boris Johnson set out to redraw the country's cultural and economic map. More of a slogan and less of a coherent set of policies, the Conservatives committed to the 'levelling up' of Britain's regions. The 2019 Conservative manifesto put forward a 'levelling up agenda' aiming at regional investment, devolution of administrative power and reskilling the country's workforce with an expanded programme of apprenticeships (Conservative and Unionist Party, 2019: 26, 36).

Thus far, and in terms of policy implementation, 'levelling up' has been used by government officials as a catch-all phrase for criticising a London-centric economic and cultural model and expressing a desire to increase central spending on regional infrastructure in left behind regions. However, this is not the first time that Conservative politicians have declared their intentions to address and rectify regional disparities and grievances. The Northern Powerhouse (HM Treasury, 2016) is a set of proposals aiming at the economic development of the North of England by the 2010–2015 Conservative-led coalition government. Similar to the 'levelling up' agenda, the Northern Powerhouse functioned as a means to decentralise the British economy and to restore a sense of pride and economic importance to the country's northern cities and towns. The proposal involved improvement in transport links, investment in science and innovation, and devolution of administrative powers. In order to avoid any misunderstandings with regard to the role of the state in the market economy, successive Conservative governments emphasised that these rebalancing plans are driven by the 'flourishing' private sector and supported by 'innovative' local governments (HM Treasury, 2016) and that the objective of the increased funding in research and development is to 'attract and kickstart' private investment.

Talks, plans and policies on reinvigorating the national economy and reforming the competitive labour market avoid in a careful and systematic way the topic of class, wealth and income inequalities. Instead, the objective of the 'levelling up' and the 'Northern Powerhouse' agendas is to ensure all parts and regions of the country productively participate in capitalism and benefit from the expected economic growth. This approach can be explained by two different yet interconnected ideological and administrative developments. First, it is important to note that despite the potency of the message of the 'levelling up' agenda, the government neither suggests a programme of wealth redistribution through progressive taxation nor denies that inequality has no productive role to play in the economy and society. The 'levelling up' agenda purposefully and conveniently overlooks the actual debate on inequality because it is limited to helping certain regions without addressing the gap between the rich and the poor, the secure and insecure in the country and among the regions in need of levelling up. Since Thatcher's aggressive deregulation programmes and David Cameron's austerity both debate and policies have shifted towards the creation of a level playing field of opportunity but not necessarily towards the eradication of inequality in order to avoid accusations of succumbing to the politics of envy and to the Left's 'spiteful egalitarianism'.

Second, the 'levelling up' agenda shifts the debate on inequality from the exploitation of workers in permanent and fixed-term contracts to the possible inclusion of everyone in the competitive market economy. The new dividing line in the society envisioned by the 'levelling up' agenda is not about rich and poor, secure and insecure employment but between those who have access to the market and those who in varying degrees do not.

The reluctance to legislate in favour of secure employment, higher wages and redistribution of wealth through progressive taxation derives from the chronic and persistent existence of non-standard employment and from the belief in the resilience of society and individuals to endure the volatile labour market, and potentially overcome these otherwise objective social difficulties through their own determination and ingenuity. With that in mind, it is important to highlight that secure and good employment as means to create an equal society is a relic of great governmental aspirations that never materialised. As Benanav (2020) notes, and as we show in Chapters 1 and 4, work insecurity is widespread and very few workers can claim to work in a sector immune from technological changes and the relaxation or outright dismantling of rights and protection. Yet, the association of work and social security with governmental policies and services provided by the state cannot and should not be treated as universal fact. If so, we are running the risk of ignoring older systems of social protection such as familial and communal ties. According to Andrea Komlosy (2018) these older systems of social protection have not reached their expiration date just because we live in a complex capitalist system of automated technologies and labour mobility. On the contrary, governments operating in the contemporary capitalist environment invoke and reactivate these older systems in order to substantiate austerity and the waning of the welfare state in economic and moral terms.

In this chapter we intend to focus on how inequality is both sustained and legitimised as a key aspect of the competitive labour market. Whereas in Chapter 1 we argued that the points of convergence and divergence between the discourses and policies of automation and immigration contribute to the creation of the national and xeno Homo Oeconomicus, in this chapter we explain how the topic of inequality in those two discourses is sustained and legitimised by the deployment of the concept of resilience and more specifically the resilience of individuals, societies and nations.

The chapter is comprised of three parts. The first part views the current demands for migrant labour through the prism of Agamben's conceptualisation of Aristotle's 'the use of the body'. The second part deals with automation technologies by focusing on the inequalities between humans and machines. In the third part, we argue that the explicit and necessary inequalities articulated in the discourses and policies of automation and immigration are sustained and legitimised by the hope and expectation of, and occasionally demands for, people to be resilient.

Pick for Britain? The use of the immigrant and national body

Our starting point for understanding the legitimation of inequality in the policies and discourses of immigration, and in particular the immigration of low-waged, low-skilled labour, is a philosophical inquiry into the work and use of the body. Giorgio Agamben (2015) notes that the expression 'the use of the body' is found in Aristotle's *Politics* where the nature of the slave is questioned and defined. Agamben's (2015) initial examination of Aristotle's work centres on the understanding of nature and its relationship with slavery. Even though Aristotle defines slavery as 'always contrary to nature', Agamben is quick to note that Aristotle stayed clear from dissociating the slave from other human beings. Instead, Agamben argues that Aristotle redirects the focus from questions over nature to the act of work, and further defines the slave as a human being whose work consists in the use of the body. Yet, the work of the slave is not necessarily the work of a human. It is this definition that enables Aristotle to proceed to the distinction between the work of the soul and the use of the body.

In other words, we are in a position to observe a formalised social distinction between intellectual and manual work. The use in Aristotle's expression 'the use of the body' acquires a limited understanding. More specifically, the use of the body is not meant to be understood in a productive sense but rather in a practical one. Therefore, the body of the slave for Aristotle is an indispensable component for intellectual work and has more in common with a bed or clothing than actual tools that facilitate production. Unlike other manual workers, the slave, even if s/he carries out the very same tasks as any other worker, remains a being without work mainly because her/his actions are not defined by work itself but by the use of the body. The simultaneous exclusion and inclusion from humanity provides the slave with a special status that can be explained as an act of legislation and of necessity. Agamben (2015) points out that despite the slave's exclusion from political life, her/his actual presence is an integral part of it. 'The slave in fact represents a not properly human life that renders possible for others the *bios politicos*, that is to say, the truly human life' (Agamben, 2015: 20).

Agamben's reading of Aristotle enables us to perceive the sociopolitical exclusion of certain workers as a necessary condition for flourishing economies beyond the chronological confines of antiquity. From Hegel to Keynes and Beveridge, the presence of disposable, unskilled labour with no rights and access to welfare, also known as the 'rabble', is not a historical anomaly that has or will be rectified by civil society, but rather a product of civil society's unresolved contradiction between freedom and deprivation (Mann, 2016). For Beveridge, the introduction of the welfare state will eventually dissolve this contradiction and subsequently will contribute to the re-understanding of the purpose of civil society: 'Whatever the bearing of full employment upon industrial discipline one thing is clear. A civilised community must find alternatives to starvation for preservation of industrial discipline and efficiency' (Beveridge, 1944: 300).

The consensus on the study of social and economic inequality coheres around the redistribution of wealth, tax and state expenditure on people and regions. The social and political consequences of inequality have been discussed and examined from a wide range of perspectives that usually involve class, race and gender in relation to the availability and regulation of rights, goods and services. However, what goes largely unnoticed in the respective studies of automation and immigration is that certain people might have the entitlement to command other people. Since Rousseau's *Social Contract*, inequality has acquired an expanded meaning that tests the legitimacy of waged labour and of the competitive labour market. 'By equality, we should understand, not that the degrees of power and riches are to be absolutely identical for everybody but that ... in respect of riches no citizen shall ever be wealthy enough to buy another, and none poor enough to be forced to sell himself' (Rousseau, 1991: 195). For Rousseau this type of inequality that entails the entitlement and capacity of rich people to command poor people for their benefit is best described as a relationship of domination and servitude. Rousseau's theorisation of inequality as a relationship of domination and servitude helps us to weave a common social and economic thread running through Aristotle's (via Agamben) use of the body, the need for immigrants to be occupied in specific jobs with low pay, and the dynamics of the contemporary labour market.

While it would be extreme and sociologically incorrect to draw parallels between the slave in the ancient Greek city-state and the low-skilled immigrant worker in contemporary UK, it is worth noting that the marginalisation of workers, and in particular, the permanent otherness of immigrants, have been historically beneficial to national economies concerning growth and productivity. A report by the National Institute of Economic and Social Research (Rolfe, 2016) into the impact of Brexit on immigration and employment highlights the concerns employers have over the termination of the free movement of labour. The report indicates that Brexit and the subsequent termination of free movement of labour would exacerbate the chronic recruitment problems employers in hospitality, farming and agriculture face. The concerns, according to the report, are not limited to the volume of workers coming to the UK but rather extend to their work ethos and performance: 'Everywhere you go in the leisure sector you will see lots of people from Eastern Europe. And it is not because they are any cheaper because we have the minimum wage and we have the national living wage. But they deliver a far better experience' (Rolfe, 2016: 8).

The report concludes by stating that the inflexibility of British workers and their expectations of higher wages make their recruitment almost prohibitive for future businesses in a post-Brexit economic environment.

Farmers expressed similar concerns over immigration restrictions. In 2017, fruit and vegetable farms in the UK were in need of thousands of immigrant workers and, as a result, produce were left to rot in the fields. According to survey data from the National Farmers' Union (2017), more than 4,300 vacancies were left unfilled, which approximately covers 50 per cent of the horticultural labour market. The survey points out that 99 per cent of the seasonal workers employed in the sector are from Eastern Europe and 0.6 per cent from the UK. Farmers in the UK claim that the vote to leave the EU has created the perception among foreign workers that the country is xenophobic and unwelcoming. The then Secretary of State for Environment, Food and Rural Affairs and prominent political figure in the campaign to leave the EU, Michael Gove, addressed the concerns of farmers by insisting workers from the EU would still be able to work in UK farms, but they would have to get registered and sign temporary contracts in order to 'augment' the horticultural labour force post-Brexit (Ferguson, 2017). For the Secretary of State, one way to reconcile popular demands for control of immigration with economic growth would be to import seasonal workers according to specific demands in the labour market and effectively reinstate the Seasonal Agricultural Workers Scheme. He explained that the current limited mobility of immigrant workers is not due to the negative perceptions of the UK as racist, but due to the growing economies of their countries of origin. 'It's already the case that the supply of labour from EU27 countries is diminishing as their economies recover and grow. So, in the future, we will need to look further afield' (Hughes and Daneshkhu, 2018). Yet, a survey conducted by the Royal Association of British Dairy Farmers (Grant, 2017) revealed that EU migrant workers are neither employed on a temporary basis nor are they perceived as transient workers. They are usually employed and perceived by their employers as permanent and integral members of the workforce.

As in the leisure and hospitality sectors, employers in the farming and horticultural sectors are reluctant to employ and invest in the training of workers from the local and national labour markets. Preference for immigrant over domestic workers is usually explained by the willingness of the former to work more hours, positively respond to flexible working schedules, and live in the accommodation spaces provided by farms. Yet, the reintroduction of the seasonal workers scheme in 2020, dubbed Seasonal Workers Pilot, signalled a new and regressive understanding of the meaning of immigrant labour. The government made available 30,000 visas for those who want to work in British farms for a period of up to six months but at the same time insisted that investment in automation technologies is needed in order to reduce the sector's reliance on immigrant labour. Workers from the EU who came to the UK before the end of 2020 and under the auspices of freedom of movement could move freely within the country and change jobs within and across sectors. But in the post-Brexit economy, immigrant workers can only travel to the UK as long as they have secured a job in the horticultural sector and can only change roles with the help of the agency that administered their employment contracts (Department for Environment, Food and Rural Affairs, 2020).

Reports on the state of the economy during and after the Brexit negotiations and demands from professional bodies for sustaining the current levels of immigrant labour in the UK illustrate the importance of immigration and of immigrant workers. Following Abdelmalek Sayad's (2007) remarks on the social implications of immigration, it can be argued that immigrant workers are only accepted as long as they are defined by the work they perform. For Sayad, 'the immigrant is only a body' (2007: 213) that is required to perform certain tasks and subsequently present her/himself as labour power. Outside the environments in which immigrants work and live, they are considered 'minors' who need to be taught the local customs and the demands of the host economy and society. Sayad (2007) substantiates his argument on the singular perception of the immigrant as a working body by distinguishing between body and head, between working and thinking.

Here, the implicit references to Aristotle's 'use of the body' for defining the nature of the slave are too strong to ignore. Low-skilled immigrant workers embody a cultural, ethnic and economic otherness that determines their position in both economy and society. They usually come from poorer countries and are part of different social and historical processes, which are sometime incompatible with the host country's social and political order. Up to a certain extent, the otherness and incompatibility that low-skilled immigrants embody are desirable qualities for the justification of low wages as well as for the regulation of their rights and mobility. They find themselves at the bottom of the social and political order due to their treatment by their employers and policymakers as convenient and advantageous work accessories.

The attempt of the UK's Conservative government to tame anti-immigration sentiments with plans to set salary thresholds and prioritise the inflow of high-skilled immigrants over low-paid and low-skilled immigrants has been refuted by businesses in agriculture, farming, hospitality and leisure. Low wages, the willingness to do jobs British workers are unwilling to do in conjunction with the objective difficulties to form or be part of a trade union constitute low-skilled immigrant workers as vital components for the country's economic growth and productivity.

Would it be possible for a government and in particular a Conservative government elected on the promise to cut down immigration to reconcile anti-immigration sentiments with the needs of businesses to have access to low-paid, low-skilled immigrant labour? Inspired by a Second World War campaign and motivated by a renewed interest in people and regions removed from the financial sector and service economy, the UK government launched a portal and a nationwide campaign titled Pick for Britain (Pick for Britain, 2020). For decades, the UK's agriculture has been reliant on immigrant labour. Every year, approximately 80,000 workers, mainly from Bulgaria, Romania and Lithuania, come to harvest the UK's fruit and vegetables (Russell, 2020; Adkins, 2020). But in 2020 the combined problems of the COVID-19 pandemic and the ongoing uncertainties over labour mobility, the right to work and stay caused by Brexit prevented many fruit pickers from making the journey to the UK. This left a significant gap in the UK's agricultural workforce that had to be filled by British workers.

With the support of the NUF and the agricultural recruitment agencies Concordia and HOPS, the Secretary of State for Environment, Food and Rural Affairs George Eustice estimated that 'probably only about a third of the people that would normally come here are already here, and small numbers may continue to rise' (Adkins, 2020). Even though the campaign aimed at addressing rising unemployment and lack of immigrant workers due to the COVID-19 pandemic and Brexit respectively, the major advocates of Pick for Britain stressed the harsh working conditions and low pay this type of work entails. The chief executive of the recruiting agency Concordia stated that the newly recruited pickers should expect to earn an average of £350 a week for approximately 40 hours of work. In addition to low pay, both the government and the recruitment agencies could not hide the fact that the working and living conditions are demanding and harsh. Work starts at 5am and finishes in the afternoon in order to avoid working during the summer's afternoon heat. Most seasonal workers, British and foreign, would be staying in on-site accommodation provided by the employers and a charge of £50 per worker sharing accommodation with two other workers is subtracted from their total weekly pay.

The campaign to recruit workers in jobs with low pay and in harsh conditions had to be accompanied by a message of duty, patriotism and hope. Not only did the government conspicuously paraphrase the famous Dig for Victory campaign that urged Britons to grow their own food during the Second World War but more importantly it established parallels between the war and the COVID-19 crisis and ascribed a sense of agency to British people throughout the country. In order to ensure a positive outcome for the campaign and avoid 'a disastrous situation in which mountains of food are left to rot' (Adkins, 2020), the UK Conservative government deployed Prince Charles as the principal messenger of patriotic duty, sacrifice and of the Blitz spirit to urge British citizens to undertake unglamorous and challenging jobs for the good of their country.

The call to undertake challenging and unglamorous jobs previously advertised to immigrant workers problematises both the use of the immigrant body as an instrument to economic growth and notions of inequality based on national categories and hierarchies. Yet, the contextualisation of Brexit uncertainties and of the pandemic as unprecedented national crises have enabled the government to invoke experiences and historical accounts of national unity and sacrifice, and effectively reverse the national hierarchy of labour and domination.

'Heroes' working with and for robots

While the policies, strategies and debates on the levelling up agenda and Pick for Britain provide a clear understanding of the contemporary components and social dynamics of inequality with respect to regional productivity and low-skilled, low-paid labour, the debate on automation and inequality is more complicated. How can we approach the inequality generated and sustained by the automation of work? An important empirical and theoretical problem with automation's impact on inequality is the former's ambition to redraw the labour landscape and render class and class politics obsolete. It is important to stress that there is a considerable discrepancy between ambition and lived reality, but at the same time we cannot ignore the gradual waning of the meaning of work for understanding social inequalities. As early as in 1982 Andre Gorz noted that 'any employment seems to be accidental and provisional, every type of work purely contingent' (Gorz, 1982: 70). In contrast to Marx's definition of the proletariat and of the working classes as the class of waged workers engaged in industrial production and to all people who must work for their living and who receive wages or a salary in technical and service professions, the new working classes neither define themselves in relation to their employment nor to their position within the social process of production. The question of how to categorise a university lecturer on a fixed contract, a freelance journalist, a care worker, a software engineer and a lorry driver is no longer as important as it used to be if we consider the growing insecurity even among the most secure professions and the numerical majority of people expected to work across industry and trade sectors.

Even though retraining programmes and the acquisition of transferable employability skills aim at maximum productivity and the smooth transition from one job to another, they create a permanent uncertainty and confusion over the meaning of work. The only certainty, Gorz (1982) points out, is that workers do not feel they belong to any of the existing classes as defined by economists, demographers and sociologists. Consequently, work ceases to be an individual contributing factor to the total output of any given national economy and society. Instead, the total output is prioritised over individual contribution and work becomes the product of the machinery of social and national production. Put differently, Gorz (1982) illustrates the process in which work ceases to be the property of individuals, it remains external to them, and now belongs to 'the machinery of social production'. Following through this process, not only robots have stolen work from individuals but have also transformed it into a mode of subordination, and as Rousseau (1991) would say into a mode of domination and servitude.

Gorz's observations on the disappearance of a recognisable class order concerning the working classes and on the way the machinery of social production both as a concept and output precedes individual work enables us to shift the focus from inequality between regions, national citizens and immigrant workers to inequality between robots and human workers. The surge in demand for online shopping as a result of repetitive lockdowns during the COVID-19 pandemic accelerated the process of automating warehouses and at the same time categorised warehouse workers as essential workers.

The paradox of the rise of robots in warehouses and the categorisation of warehouse workers as essential was encapsulated in a series of Amazon advertisements titled Thank you Amazon Teams, each one focusing on a specific country in which Amazon maintains a business presence (Amazon

Stories EU, 2020). The UK advertisement begins with an aerial view of an Amazon warehouse, and then the viewer is taken inside the well organised warehouse where workers and machines such as automated forklifts and trolleys, scanners and conveyor belts work side by side in harmony. A voiceover stresses the importance of Amazon during the pandemic and praises its workers: "Right now, delivering the things people need has never been more important. To all our Amazon team on the floor, and on the road – thank you" (Amazon Stories EU, 2020). Amazon's public appraisal of its workers disguises the fact that in many respects the latter perform a minor role compared to the warehouse robots. The robotisation of warehouses was managerially and technologically informed by the electronic tagging of human workers. In 2018, Amazon approved and started using wristband trackers transmitting data for monitoring the (human) picker's hand in relation to the warehouse's inventory in order to provide feedback to managers and motivate workers to improve their performance. Consequently, the digital monitoring and quantification of the warehouse worker is justified by proclamations of workers' efficiency as well as wellbeing.

Statista predicts the global warehouse automation market will increase from $15bn in 2017 to $30bn by 2026 (Mazaranau, 2020). Yet, robots are not necessarily replacing human workers in warehouses. Many robots cannot compete with the efficiency and handling skills of human hands and as a result most warehouse jobs are becoming a hybrid of human–robot interaction and alleged collaboration. Chuck is an autonomous robot trolley that is programmed to lead human workers known as 'pickers' to designated shelving units (6 River Systems, nd). The company that provides robots to big warehouse operators such as DHL and Office Depot states that new technological advancements in the organisation and division of labour contribute to better and easier working conditions for the workers because they do not have to push trolleys around vast warehouses. Yet, human workers are unable to keep up with Chuck's pace. According to 6 River Systems, human workers slow down and underperform without robotic assistance.

Amazon's robotic services, Kiva, follow a different approach to human–robot collaboration. In the automated warehouses of Amazon, robots bring the shelves to human workers in order to minimise unnecessary mobility and contribute to better working conditions (Amazon Robotics, nd). Again, for Kiva, renamed as Amazon Robotics, the main objective of automated services is the rise of human productivity. Working without robotic assistance, the average worker picks 100 items per hour but working with robotic assistance, the average worker picks 300 items per hour. Whereas lack of mobility becomes synonymous with higher productivity and efficiency, Amazon's rollout of automation programmes failed to foresee the impact of stationary work on workers' health. Chronic pain and various orthopaedic problems are the unintended consequence of Amazon's heroes working with and for warehouse robots.

Resilience as the answer to inequality and to the failings of the competitive labour market

Most research on market economies focuses on the neoliberal remaking of the state and in particular on the gradual or at times aggressive transformation of key institutions and services usually and up until recently managed by elected national governments (Dardot and Laval, 2013; Brown, 2019). But the overwhelming emphasis on the state and its remaking according to business standards and activities neglects another parallel and equally important transformation; that of the remaking of businesses according to state standards and activities. For Vauchez and France (2020), this blind spot in the analysis of neoliberalism can be attributed to traditional and persistent notions which tend to treat the state and the private sector as opponents fighting for their territory in the market economy. In what Chamayou (2021) terms 'authoritarian liberalism' for describing the enhanced role of corporations and managerial rule of society, the economic order of competition and the free market is neither maintained nor reproduced in a spontaneous and natural manner. On the contrary, the state and corporations have to converge, borrow from each other's slogans, administrative and managerial strategies and present themselves as a unifying and coherent entity whose purpose is to present the economy as a depoliticised domain operating under the misapprehension of 'horizontal comradeship' (Anderson, 1991). Although Benedict Anderson (1991) argued that the defining quality of this type of comradeship is the willingness of the members of the imagined communities, better known as national subjects, to die for this community, the horizontal comradeship communicated by the political and economic nexus of the nation and the corporation elevates work as the ultimate contribution to society.

The campaigns Pick for Britain and Thank you Amazon Teams indicate that the territorial battles between state and corporations are long gone and that corporations have to address their employees as dutiful citizens whose work is essential for the survival of both the nation and of capitalism. The categorisation of warehouse workers as heroic and the appeals to national duty as part of the Pick for Britain campaign are grouped together and meant to be understood through the concept of resilience. Resilience aims to contribute to a sociopolitical regime in which poor working conditions, low pay and inequality are justified, even endorsed, for the greater good of lowering immigration and growing the economy during the pandemic. Joseph (2013) points out that the intellectual roots of resilience can be found in the ecology literature dealing with the impact of global environmental change but concerns over externally generated threats can easily be applied to the domains of politics and society. This analogy enables us to examine the capacity of institutions, of legal and policy frameworks to absorb the shocks and adapt to changes created by the crises of Brexit and the COVID-19 pandemic respectively.

Resilience enters the sociological and political lexicon as soon as there is the realisation that complexity, precarity and contingency dominate all aspects of individual and public life. Such a realisation and, dare we say, acceptance of precarious working conditions, low pay, and inequalities between north and south, national citizens and immigrants, robots and human workers allow a limited space for meaningful change and deprive individuals of the possibility of a meaningful intervention in their work environment. The only viable option emanating from the government's Pick for Britain campaign and Amazon's appraisal of its workers whose tasks and performance targets are dictated by robots is resilience – in other words to learn how to absorb the shocks of the competitive labour market and adapt to the normality of a decrease in immigration flows, labour shortages, low pay and the dominating presence of robots in the working environment.

It is precisely because of this shift from collective to individual responsibility and the prevailing certainty that exposure to precarity, inequality and contingency are unavoidable if not necessary components of the competitive labour market, that resilience has been associated with the logic of neoliberalism (Joseph, 2013; Zebrowski, 2013). In the first instance, resilience appears to be an all-encompassing concept as well as strategy for coping with the uncertainty, insecurity and occasional opportunities evident in the discourse of neoliberalism and in the policy framework surrounding the competitive labour market. The implicit or explicit deployment of resilience in the Pick for Britain and Thank you Amazon Teams campaigns amplifies the need for, and moral obligation of, reflexivity, adaptability and sacrifice. At the same time, these expectations or outright demands not only disqualify any criticism of the present and future direction of economy and society but in a more specific way they also suggest the impossibility of higher salaries and better working conditions.

The widening gap between the demands and aspirations of the competitive labour market and the inability of workers to participate in it has contributed to the gradual theoretical and empirical detachment of resilience from the logic of neoliberalism (Chandler, 2014; Schmidt, 2015). Rather than being a manifestation of neoliberalism and an effective mechanism for coping with the uncertainties and risks of the market economy, Schmidt argues that 'resilience does not operate in a continuation of a paramount neoliberal government paradigm but can be understood as a response to its inherent frustrations' (Schmidt, 2015: 404). On a similar tone, Chandler (2014) points out that governance can no longer be conceived as a set of bottom-up interventions for incentivising workers to participate in the market, but more accurately as a transformative process that ascribes an agency to resilient workers. For Chandler (2014), the workers' adaptation to changes, shocks and disruptions should not be perceived as an indication of hopelessness and complacency but as an expression of agency negotiated and manifested in complex economic, political and social environments.

It is important to acknowledge that resilience both as a concept and a coping mechanism can be associated with and utilised by neoliberalism and post-neoliberalism respectively. However, both sets of analyses overemphasise the importance of uncertainty and lack of regulation in competitive market economies at the expense of a controlling set of policies and management techniques whose sole purpose is to disguise the failures and inconsistencies of the competitive labour market. The campaigns Pick for Britain and Thank you Amazon Teams deploy resilience in a manner that is at once neoliberal and post-neoliberal but more importantly as an authoritarian managerial instrument that disguises the failures of labour market; denies any sense of control of the market; and addresses adaptability as a corporate-patriotic duty and obligation. The good of the nation and the good of the corporation converge in the interpellation of the worker as a subject that needs to correspond and adapt to demands for lower immigration and increasing volume of online retail while accepting inequality concerning salaries and working conditions as the inevitability of a national economy incapable of reconciling immigration with the free market and automation with decent working conditions.

4

I, Robot

The machine and the worker

What do we talk about when we talk about automation? Capitalism's continuous drive towards destruction and renewal is generally accepted as an economic and philosophical platitude with no real desire for contestation. The historisation of capitalism usually involves the identification of 'stages' that correspond to specific historical experiences. Since Marx's famous dictum in *The Poverty of Philosophy* that 'the hand-mill gives you society with the feudal lord; the steam-mill society with the industrial capitalist', the relationship between productive forces, modes of production, earnings and social relations is quintessentially a relationship framed and structured by the overarching economic and political system – namely capitalism. Perhaps, one of the most important and long-lasting contributions of Marx's observations in *The Poverty of Philosophy* is the social embeddedness of technology. For Marx, technology is neither an objective resource nor a socially detached scientific innovation, but instead a vital co-constitutive component of society and economy and a vital instrument for the reproduction of social inequalities. Against Proudhon's optimistic assertion that the machine is 'the logical antithesis of the division of labour', Marx postulates that the machine is the unification of the instruments of labour and by no means does it constitute a combination of the different operations the worker has to perform. Far from signalling the end of the division of labour, Marx argues that the introduction of machinery has exacerbated the division of labour in society, rendered the task of the worker inside the workspace as simplistic, and contributed to the concentration of capital.

However, Marx's writings on machines and technology were never one-dimensional. Eight years before the publication of *The Poverty of Philosophy* (1976 [1847]), Marx's *Grundrisse* (1993 [1839]) offered a glimpse of the major themes explored in *Capital* but also provided a rare opportunity to envision an economy where labour in all its manifestations such as human, animal, manual and cognitive, has been replaced with machines. Due to its rather late publication in the Soviet Union in 1939, in West Germany in 1959 and in the UK in 1973, the book never reached the popularity of *The Communist Manifesto* (Marx and Engels, 1967 [1848]) and of *Capital* (2004 [1867]). While *Grundrisse* as a collection of texts, or to be more precise notebooks, had at the time a limited impact, a specific section titled 'The fragment on machines' has served as an economic and sociological springboard for examining automation, robotics and artificial intelligence (AI). In that fragment, Marx postulated that due to the accumulation of scientific knowledge, described as 'general intellect', labour will play a lesser role to capitalist accumulation, effectively causing a crisis of the labour theory of value as illustrated not only by Marx himself but also by Adam Smith and David Ricardo. According to Marx, as long as labour retains its conventional meaning, then:

> directly, historically, adopted by capital and included in its realisation process, it undergoes a merely formal modification, by appearing now as a means of labour not only in regard to its material side, but also at the same time as a particular mode of presence of capital determined by its process – as *fixed capital*. (1993 [1839]: 693, emphasis original)

At the same time, Marx is quick to note that the production process is no longer a process dominated by labour as such but instead labour is incorporated in the wider process of machinery. The transformation of the means of labour into technological equipment and of human labour into an accessory of this equipment not only creates an uneven relationship between machine and human but most importantly addresses knowledge as capital's last frontier. Marx's pessimistic tone becomes clear when he considers the accumulation of all knowledge and skills relevant to the production and operation of machines is absorbed into capital as opposed to labour, and as a result knowledge appears as an attribute of capital. In the age of machinery, rail and telegraph networks Marx raises questions over the value of knowledge and science, and their role in capitalist accumulation. The 'Fragment on machines' considers knowledge in three different yet interrelated ways: first, knowledge is a direct force of production; second, knowledge is a social force of production; and third, knowledge becomes social practice. But if this thesis is taken to its logical conclusion, there should not be any reason to be pessimistic about living labour's relevance and power. Machines and knowledge do not only expand industrial production but also decrease physical effort. As a result, machines will increase free time and eventually liberate humans from wage labour. Such a result can only amount to a capitalist crisis in which workers freed from the physical constraints of labour will grow both collectively and intellectually.

Two different economic and social processes rekindled the interest in Marx's theoretical observations on machines and labour. The Italian *operaismo* movement via the work of Paulo Virno (2004) and Mario Tronti (2019), among others, interpreted the 'Fragment on machines' as a philosophical response to the supposed neutrality of science at the service of industrial production. Situating the factory at the centre of social and economic life, *operaismo* drew the conclusion that industrial relations do not only reflect social and political life, but most importantly determine it. But it was the process of post-industrialisation and the advent of service and knowledge economies that enabled what became known as post-*operaismo* to solidify the 'Fragment on machines' in the struggle against capitalism.

Michael Hardt and Toni Negri's (2001) *Empire* advanced the arguments of *operaismo* and at the same time developed a polemic response to global capitalism, automation and interconnectedness. Echoing Marx's fragment, they argue that one of the most significant developments in globalisation's latest phase is the presence of 'immaterial labour' defined as a type of labour that surpasses 'the expropriation of value measured by individual or collective labour time'. Once again, capitalism contains the seeds of its own destruction. In the *Empire*, immaterial labour is beyond capital's control because the former coheres and organises around the multitude. Drawing on Spinoza, Hardt and Negri's reformulation of the concept of multitude signals the end of the male, White manual worker and the emergence of a 'mobile body of so-called singularities' (Hardt and Negri, 2001: 53). The multitude's otherwise immeasurable productivity can be found in the communicative and affective networks of global capitalism resulting in labour's capacity to define and measure its own value irrespective of capitalist metric systems. As knowledge and cognitive skills escape capital's supervision, they become direct producers of value. For Hardt and Negri (2001) as well as for Virno (2004) this the most accurate manifestation of Marx's 'general intellect'.

Discussing automation and technological progress: economic growth, inequality and labour fragmentation

The optimistic theoretical accounts of labour asserting its own value within capitalism's communicative and affective networks have been met or at points tamed by a series of policies, reports and government directives arguing for the re-evaluation of knowledge and skills in an automated economy. There are two outstanding issues that governments and corporations have to deal with: how knowledge, cognitive and interpersonal skills will serve a growing and productive economy; and how to mitigate the negative repercussions of an automated economy, namely mass unemployment and retraining of the current labour force.

A conventional and not necessarily incorrect way to approach the current debate on automation would be to identify the techno-optimists and the techno-pessimists and discuss their respective positions. In that respect, techno-optimists are of the view that automation should not only be encouraged but more importantly accelerated for the purposes of achieving higher standards of living and reconfiguring the notion of work in both economy and society. On the other hand, techno-pessimists are concerned with the negative repercussions of automation and in particular with mass joblessness and rising inequality. While the distinction between techno-pessimists and optimists might be helpful for mapping the debate on automation, its analytic qualities are limited due its inability to grasp political and ideological complexities, and the context in which the debate takes place. Both techno-optimism and pessimism can exist as legitimate positions for either low taxes and free markets or for an economic model that advocates regulation and wealth redistribution. Yet, the current debate on automation is full of false starts and paradoxes. Automation is perceived and deployed by policymakers as the answer to stagnant economic growth and chronic low productivity, but governments thus far are not willing to distance themselves from high employment figures nor from the conventions of paid jobs.

Political and economic tensions between labour and technology have been a permanent feature of capitalism. Karl Marx argued in *Capital* that such tensions are the direct outcome of the continuous development of the forces of production. The technological advancements of industrialisation and automation expose the contradictions between productivity and the fear of being excluded from the production processes of goods and services. Human labour has always been under threat by industrialisation, automation and more recently by robotics and AI. However, this threat has been extended to questions over the present and future state of capitalism as a sociopolitical order. According to Streeck (2016), capitalism's existential crisis is illustrated by two distinctive yet interrelated phenomena: first, the process of replacing human labour is steadfast and largely unopposed due to lagging productivity and stagnant economic growth; second, the target of automation has now shifted from the manual working class to the service-oriented middle-class professionals – in other words to the apparent defenders of an economic system based on private property, free market oriented economy and entrepreneurial activity. Robots and automation might render a significant percentage of jobs obsolete, but their sole purpose is to minimise taxation, increase productivity and maximise profits.

Does contemporary capitalism demarcate new battle lines between workers and robots? Fears in public and political discourses have been supported and legitimised by studies claiming massive loss of jobs and urgently demanding the re-evaluation of existing and future skills. The study that set the mood over such fears was the 2013 study by Carl Frey and Michael Osborne of the Oxford Martin School. Motivated by John Maynard Keynes' infamous prediction of widespread technological unemployment 'due to our discovery of means of economising the use of labour outrunning the pace at which we can find new uses of labour' (Keynes, 2015 [1930]: 80), Frey and Osborne pursued a quantitative analysis of the meaning and effects of technology in relation to the future of employment. The study predicted that 47 per cent of jobs in the US were at risk due to automation. Following the impact of their predictions, the *BBC News* website translated the study into an accessible platform that enabled the public to check if their current profession is susceptible to the advent of automation. Frey and Osborne distinguish between high, medium and low risk professions on their probability of computerisation and highlight transportation and logistics, manufacturing, office and administrative jobs at risk. One of the most surprising findings of this study is that a substantial percentage of jobs in the service sector are susceptible to computerisation. The recent growth of service robots at once signals the end of the assumed advantage of human labour involving flexibility and sociability and challenges the recent drive towards service jobs as a means to tackle unemployment in the advent of deindustrialisation and automation. One of the starkest predictions of Frey and Osborne's study is the development of a potential negative relationship between educational qualifications and wages. They claim that the process of automation and computerisation will not be limited to manual low-paid jobs but instead will cause the gradual disappearance of middle-income jobs too.

Soon after the publication of Frey and Osborne's study, think tanks, accounting and financial services firms attempted to ease fears of joblessness by making the case for the positive impact of automation on productivity as well as on the composition of the competitive labour market. Deloitte's 2015 report titled *From Brawn to Brains: The Impact of Technology on Jobs in the UK* aimed at creating a better understanding of the 'actual' effects of technology on the UK's present and future workforce compared to previous hypothetical studies. Deloitte's (2015) research found that over the last 15 years the UK has benefited from the recent surge in automation, computerisation and robotics. While the study does not deviate from the conventional narrative of massive job losses, it points out that 3.5 million jobs have been created by the advent of automation and robotics and 800,000 have been lost. The jobs growth reported by Deloitte is accompanied by a positive outlook concerning salary increases. According to the report, on average, each job created is paid £10,000 per annum more than the low-skilled, routine jobs it replaces and as a result contributes more to the economy across Britain's regions. Deloitte's report indicates that this trend can only be sustainable if future businesses have access to people with a new and distinctive set of skills: digital know-how, management capability, creativity, entrepreneurship and complex problem solving. The report revisits and subsequently reaffirms Frey and Osborne's (2013) 'hollowing out' assertion.

Routine cognitive jobs that require precision and numeracy skills are not performed by low-skilled workers, but such jobs are increasingly susceptible to the technological advances in big data and machine learning. However, Deloitte's report deviates from Frey and Osborne's 'hollowing out' assertion by differentiating between tasks and occupations. Technology, the report argues, can have a great impact on specific tasks and by association on the character of the occupation without necessarily eliminating the actual occupation from the social and economic landscape. The optimistic tone of the report derives from the refusal to accept automation as a zero-sum game. Humans and machines can develop a symbiotic relationship in which human intelligence and work performance are enhanced instead of being threatened or substituted.

Contrary to the rather optimistic and conciliatory tone of Deloitte's report, in 2016 the World Economic Forum warned that 'five million jobs will be lost by 2020'. According to the Forum's *Future of Jobs* report (2016) skill-sets will have to be radically different in order to keep pace with the demands and disruptions of the 'fourth industrial revolution'. Education, retraining and the diversity of workforce are of paramount importance for surviving and thriving in a highly competitive environment. One of the most interesting aspects of this report is the expectation that businesses should take the initiative to retrain their staff for learning new skills. Such an expectation does not limit itself to issues surrounding employment figures and staff welfare but more importantly to the sustainability and wider relevance of businesses. It is worth highlighting that the World Economic Forum acknowledges that the 'fourth industrial revolution' disregards national borders as well as physical and organisational spaces and while training and reskilling remain essential businesses could and should source their workforce from across the world.

The driverless car is one of the most indicative examples for communicating anxieties about labour displacement and a sense of optimism about a productive, growing economy respectively. If cars can drive themselves, what will happen to professional drivers? Will they be permanently displaced by technological advancements? The UK government's 2017 Autumn Budget illustrates the need for more automation and autonomous technology in order to tackle chronic low productivity. Relying on figures from the Organisation for Economic Co-operation and Development (OECD), the budget states that productivity growth among developed economies has been at least 1 percentage point slower since 2008 than in the preceding decade. In the UK, according to the figures provided in the budget, the slowdown has been more severe and productivity growth has averaged 0.1 per cent since 2008 compared to 2.1 per cent in the preceding decade (HM Treasury, 2017: 11). One of the main arguments put forward in the Autumn Budget of 2017 was that 'an economy driven by innovation will place the UK as a world leader in new technologies such as Artificial Intelligence (AI), immersive technology, driverless cars, life sciences and Fin Tech' (HM Treasury, 2017: 43). A significant part of this new technology and innovation agenda deals with what the government calls 'connected and autonomous vehicles'. The government expressed the ambition to see autonomous, self-driving cars on UK roads by 2021. This ambition will be realised through changes in the regulatory framework for testing autonomous vehicles without human operators, and through the adjustment of present and future road networks to support these vehicles. In a subsequent interview, the then Chancellor of the Exchequer Philip Hammond predicted that a million British workers would have to retrain and acquire a new set of professional skills in order to be employed in new jobs. "It is going to revolutionise our lives; it is going to revolutionise the way we work. And for some people, this will be very challenging" ('Today Programme', 2017).

The ambition to populate roads with driverless cars extends to the transportation of goods. While corporations such as Google and Toyota have been active in the development of cars, which will revolutionise work and everyday life, lorry manufacturers are developing similar technologies aiming at minimising costs in transporting.goods. 'Labour accounts for up to 45 per cent of total road freight cost' (International Transport Forum, 2017: 9) and driverless lorries will redefine profit margins and ultimately redefine the rules of competition. As a result, up to 4.4 million of the 6.4 million professional lorry driver jobs in the US and the Europe could be eliminated by autonomous technology (International Transport Forum, 2017: 7). Even though the report praises the resilience and adaptability of the workforce under consideration, it concludes that their relative low education level would be almost prohibitive for acquiring new and diverse skills and knowledge.

The links between automation, education and labour displacement are further explored in a report published by PricewaterhouseCoopers (Berriman, 2017). The report emphasises the potential loss of jobs due to accelerated automation and predicts that 30 per cent of jobs in the UK could potentially be at high risk by the early 2030s, compared with 38 per cent in the US, 35 per cent in Germany and 21 per cent in Japan (Berriman, 2017). According to the report employment sectors such as transport and storage, manufacturing, and wholesale and retail are more exposed to the risks of automation and subsequent loss of jobs, whereas 'health and social care' appears to be immune to robotics and automation. The figures presented by the report largely depend on the educational level of workers. University-educated workers will be better equipped to adapt to a new economy and labour market, while for workers with GCSEs or no formal qualifications automation presents a clear threat to their employment prospects. The report argues that labour displacement can be managed by the proliferation of new jobs created by new technologies and by (re)training initiatives for workers in order to acquire the right skills for a challenging new labour market.

Government intervention in the domains of education and vocational training are at the forefront of the Report of the Future of Work Commission (2017). Falling real wages and increasing inequality are some of the most significant by-products of automation. The report welcomes the new technological shift in automation, robotics and AI because of its potential to decrease working hours, increase productivity and contribute to higher standards of living. Such potential can only be realised with policy intervention in the areas of 'good work'; 'skills for the future'; 'innovation'; 'corporate governance'; 'labour rights'; and 'ethics'. The report advocates a stronger and more active role for the state in the new economic landscape for assessing the quality of work workers are engage with and the equal distribution of wealth. Lifelong learning and the establishment of education trusts in conjunction with new tax rates for incentivising business to invest more in their workforce will reintroduce the state as an active actor in the management of the economy.

Social justice and the redistribution of wealth are the focal points of the Institute of Public Policy Research's (IPPR) report, *Managing Automation: Employment, Inequality and Ethics in the Digital Age* (Lawrence et al, 2017). The IPPR joins other reports in welcoming automation as a potential force for boosting productivity and reshaping the meaning of work in a variety of employment sectors. Yet, the IPPR warns that the economic gains generated by automation need to be placed in the wider context of social inequality and poverty. The primary challenge automation poses to economies and societies is not necessarily related to production but to distribution. If the benefits of automation are evenly shared societies would be better equipped to build economies based on the principles of justice, progressive redistribution of wealth, higher income and less working time. In themselves, technological advancements such as automation, robotics and AI are neither progressive nor could they guarantee a prosperous and productive future. The IPPR warns that if automation is managed poorly it could lead to a '"paradox of plenty": society would be far richer in aggregate but, for many individuals and communities, technological change could reinforce inequalities of power and reward' (Lawrence et al, 2017: 2).

The IPPR's report provides a set of three distinct yet interconnected conclusions. First, digital technologies should be adopted by all sectors of the economy, and automation needs to be encouraged by national governments in order to improve the management of companies, and facilitate an increasing role for employees concerning workloads, wages and the overall (re)distribution of economic benefits. Second, the IPPR calls for the establishment of new institutions for informing as well as regulating how automation technologies are used and for whose benefit. Third, the report advocates 'new models of collective ownership' for ensuring the meaningful and active participation of workers in their workplaces for just and productive use of technological advancements.

A follow-up report by the IPPR (Roberts et al, 2019) deals with the already established themes of automation – namely productivity, skills, redistribution of wealth, unemployment and inequality – and at the same time raises an often-neglected issue: the different effects of automation on men and women. The IPPR's analysis shows that twice as many women as men work in occupations with a high potential for automation. Yet, the report argues that automation presents an opportunity to narrow gender inequalities. More specifically, an automated and productive economy would enable higher wages in currently low-paid jobs occupied by women and new jobs will be created with more opportunities for women to be employed in jobs with better working conditions and salaries. Central to the debate around automation and gender inequality is time. One of the key promises of automation is that the deployment of robots, algorithms and AI will provide time outside of the working environment, which in turn could alleviate women of both the paid and unpaid work that many of them face. The realisation of the opportunities offered by an automated economy will not only require the acceleration of the process of automation but also the active involvement of those affected by it, especially women.

The IPPR's report puts forward four propositions for gender quality in the digital age. The first proposition makes the case for the involvement of currently and potentially affected workers in the implementation of automation in working environments. The voices and leadership of women and other workers should be central to a managed acceleration of automation. The IPPR suggests the establishment of a new social partnership body, Productivity UK, that would help firms adapting to new technologies, and ensure that women and low-paid workers are not disproportionately affected by the advent of automation. The acceleration of automation would inevitably mean a 'capital intensive' economy perpetuating existing social inequalities. The IPPR's report points out that women are half as likely as men to hold employee shares and own on average less wealth in pensions and shares.

The second proposition highlights the need for an expansion of employee ownership trusts, and for the establishment of a Citizens' Wealth Fund for ensuring equal distribution of wealth and social benefits even for people not in formal paid occupations. As specific tasks or even whole jobs disappear, new ones will emerge. The impact of automation on social equality will depend on access to relevant training and to these new jobs. The third proposition put forward by the IPPR is centred on the managed acceleration of automation. Adoption of digital technologies and automation of work should be incorporated in the Government's industrial strategy with the aim of making the UK the most digitally advanced economy in the world by 2040. The IPPR's fourth proposition is a response to the problematic relationship between human biases and use of algorithms. The alleged objectivity of machines and software needs to be questioned and the human input in engineering and design to be acknowledged. To that end, the report proposes the establishment of the Centre for Data Ethics and Innovation for the implementation of anti-discrimination measures, and a new and expansive role for the 2010 Equality Act to provide protection against discriminatory practices perpetuated by the intensive use of data and algorithms in work environments.

National governments and policymakers were slow to react to either the pessimist or optimistic predictions of the impact on automation on the economy and society. In 2019 the Business, Energy and Industrial Strategy (BEIS) Committee of the House of Commons published a report titled *Automation and the Future of Work*. The report is structured around three major themes: economic growth and productivity; demography; and workers' welfare. While previous reports were mostly preoccupied with the impact of automation and the subsequent responses from corporations and governments, the BEIS Committee not only does not share pessimistic predictions over the impact automation but also states that the problem of the UK economy is the rather limited number of robots in the workplace. Once the leading industrialised country, the UK has fallen behind due to the reluctance of corporations to adopt cutting-edge technologies and the inability of educational institutions to train the present and future workforce adequately.

Despite the bleak realisation that 'the UK missed its chance to lead on developing industrial automation' (BEIS, 2019: 3), the report points to the pivotal role of education and training for catching up. The process of a fully automated economy needs to be initiated, supported and coordinated by governments willing to support British businesses through research funding and incentives for investment. In turn, the report argues, school and university curricula are extremely narrow and need to integrate the provision of science, technology, engineering and mathematics (STEM) subjects with the creative industries where human interaction has a competitive advantage over robots.

The first main reason the UK needs to be an enthusiastic adopter of robotics and automation is the downturn in productivity. Echoing the remarks of the 2017 Autumn Budget, the UK's productivity has grown much slower since 2008 despite various attempts of successive governments to boost productivity by investing in skills, industries and infrastructure. The Committee cites the online grocery store Ocado as a success story of a company 'at the global forefront of retail automation and bespoke industrial robotics' (BEIS, 2019: 9). The Committee lists three reasons for explaining the UK's chronic low productivity: management who do not understand the potential of automation; lack of digital skills among the existing workforce; and the existence of business environments where new technological advancements cannot necessarily be implemented.

The Committee's report provides a comparative analysis in order to position the UK economy in the wider landscape of automated productive economies. China, Japan, the US, the Republic of Korea and Germany are the major buyers of robotics. In contrast, the UK accounts for only 0.6 per cent of the annual global shipments of industrial robots and 3.5 per cent of European shipments. Compared to G7 nations, the UK is far behind in deploying industrial robots, amounting to 85 robots per 100,000 workers. The Commission's report emphasises that sales of industrial robots have decreased in the period between 2014 and 2015 compared to leading industrial countries such as Germany and Japan. Throughout the report Japan serves as a case study and success story for deploying industrial robots and legislating in implementing a dynamic automated economy. Japan's position in the global rankings of automated economies not only reveals a comprehensive industrial strategy for achieving high levels of productivity and economic growth but also a technological solution to persistent demographic problems. More specifically, the report explains that Japan's population peaked in 2008 and an increasing segment of the population are beyond the prescribed age of retirement. Japan's ageing demographics in conjunction with a widespread scepticism of relaxed immigration policies have forced successive Japanese governments to focus on robotics and automation in order to ensure labour shortages across all sectors of the economy are managed effectively and cultural and social cohesion is maintained.

But why is the UK so far behind in adopting automation? In the first instance, the austerity policies in the aftermath of the 2008 global financial crisis had a negative impact on investment in research, skills and new technologies (Clarke et al, 2017). In the second instance, the persistence of low-paid and low-skilled work acts as an obstacle for investing in automation and consequently in accepting its transformative merits. Faced with the cost and uncertainty of automation, governments and corporations are more comfortable with the reliable and cheaper human labour mostly generated by immigration. The reindustrialisation of the UK and the desired boost in productivity and economic growth can only be achieved if there is a fertile economic and political environment that allows the proliferation of robotics and automation. As long as workers perform jobs that are poorly paid because high-skilled and high-paid jobs are not available, both the government and firms are not particularly incentivised to invest in new technologies and education programmes. As Smith (2021: 11) contends, 'this prevalence of cheap labour, indexed by decades of stagnant wages for workers, is itself an effect of technological stagnation'. Despite the inability or unwillingness to implement policies and encourage the rollout of an automated economy that does not rely on immigrant labour and low wages, the discourse of automation continues to serve, among policymakers, as a medium for the projection of a technocratic and meritocratic economy that is growing and is in sync with the national as well as global demands.

Contrary to popular economic arguments expressed in favour of taxing robots (Delaney, 2017), the House of Commons report argues against robot taxation because there are not enough robots in the UK economy and this type of taxation would be ineffective and 'perverse'. At the same time, the report highlights that the cost of robotics and automation has decreased dramatically, and all corporations should be encouraged to invest in automation and robotics via research and development tax credits and capital allowances.

The House of Commons report is not limited to the economic benefits of productivity and growth but extends to the more contentious issues of job security, employability and retraining for acquiring new skills. In line with other reports, the early dystopian signals of the massive disappearance of jobs are counterbalanced by the more detailed and nuanced assertion that automation would not eliminate jobs but only tasks linked to specific jobs. One of the most innovative suggestions of the report is the creative and harmonious co-existence between workers and robots. Effectively, robots will become 'cobots' – collaborative robots with the capacity to make jobs easier while creating new jobs and opportunities for high-skilled workers.

Fears and hopes regarding the direction of the economy and the labour market are mostly structured around national economies and constellations of nations glued together by geographical proximity and shared political and economic systems. Competitiveness, efficiency, demands for ever higher economic growth and productivity dominate the debate on automation and robotics but also point to existing and potentially new hierarchies. The EU's response to the 'digital revolution' via the EU Commission's working paper titled 'The impact of technological innovation on the future of work' focuses on the consequences of technological innovations on labour markets and discusses the future direction of policy at a pan-European level regarding employment and skills. The Commission's working paper challenges the optimistic assumption that technological enhanced productivity would be beneficial to both high- and low-skilled workers. The working paper states that automation and robotics not only affect the actual number of jobs available in the labour market but also the skills needed for performing new tasks and jobs.

Technological advancements continue to favour cognitive skills in such a way that university graduates still face quite stable although no increasing returns to their education. However, the EU Commission's working paper points out that labour market prospects seem to increasingly rely on the possession of skills acquired beyond the parameters of formal education. Evidence of this observation is provided by the research of Goos et al (2019) who suggest that 'digitalisation might actually be biased not only with respect to cognitive skills that are typically obtained by education but also with personality traits' (2019: 11). Despite the presumption that workers with positive personality traits will be better equipped at defending themselves from an irreversible technological displacement, the presence of the 'cobot' in research papers and public debates signals a new era for working conditions and working relations. The 'cobot' is placed in working environments for facilitating an atypical competition between human labour and robots aiming at higher productivity and profits.

On a policy level the EU Commission's working paper coheres around two topics: first, it highlights the importance of education for mitigating the negative effects of technological displacement and equipping current and future workers with adequate skills for coping with new technologies; second, social protection is paramount in order to control existing and emerging working conditions that tend to prioritise profit over workers' welfare and living wages. The working paper points to the EU Commission's Digital Education Act Plan (European Commission, 2018) and its subsequent potential to protect unskilled workers from automation. Contemporaneously, governments are starting to act to protect workers who are being affected by regressive working conditions. The Netherlands and Finland have attempted to regulate zero-hours contracts; French workers working in and for digital platforms now enjoy full rights to set up new trade unions or participate in existing ones. Furthermore, the advent of automation has pushed governments to widen access to social protection programmes and provisions. Examples include Denmark's unemployment insurance scheme for the self-employed and workers in non-standard jobs; Italy's Jobs Act for the protection of workers in the 'smart economy'; France's principle of social liability of collaborative platforms and the latter's de facto participation in work accidents insurance coverage; and Croatia's tax reform for paying social security contributions to workers in the creative industries.

Regulation and workers' protection are the main foci of the European Trade Union Institute (ETUI; Castillo and Ponce, 2020). The institute acknowledges the disruptive effects of automation, robotics and AI but insists that a future regulatory framework needs to be in line with European values. While the meaning of such values remains vague in this Foresight Brief, the Institute believes that protection of workers is not antithetical to productivity and economic growth but instead integral to maintaining a democratic identity in an automated economy. Robotics and automation can have an impact on automation in diverse ways: trackers for Uber drivers and Deliveroo riders; nurses connected with tablets; technicians and factory workers working with robots; algorithms deciding who gets employed and promoted. For the ETUI the existing guidelines concerning the use and effects of emerging technologies are too broad and tend to focus on abstract generalisations. The development of a robust regulatory strategy on emerging technologies involves dealing with different yet intertwined aspects of the law, from the transparent and accountable design of software and artefacts to civil law liability rules for determining who is responsible and liable for the development and application of specific technologies.

To that effect, the ETUI put forward seven 'dimensions' to future regulation as points of discussion considering present and future negotiations and employers. The first dimension points to the need to ensure that workers know how to protect their privacy when working with robots, automation technologies and AI. There must be new provisions on how data on productivity and performance is extracted, analysed and for what ends. The second dimension entails issues surrounding the ownership of data and the right of employers to surveil their employees. The ETUI lists two indicative legal cases in which the European Court of Human Rights has ruled against an employer who had prohibited all personal use of IT equipment and therefore dismissed an employee who breached that rule (*Barbulesku* v *Romania*, 2017, cited in Castillo and Ponce, 2020); and in favour of two academics claiming their employer had breached their right to privacy by installing cameras in teaching spaces (*Antovic and Markovic* v *Montenegro*, 2017, cited in Castillo and Ponce, 2020).

The third dimension addresses fairness and transparency. The ETUI argues that algorithms should not only be transparent but fair to all stakeholders, namely employees involved in the productive and profitable use of technologies. Fairness implies the design of algorithms while taking into account the trade-offs between higher productivity and workers' rights, and the effects of algorithms on existing social inequalities based on race, class, gender and age. The fourth dimension involves the obligation of employers to explain decisions made by algorithms. For the ETUI, it is important to integrate 'the right to explanation' in negotiations with employers in order to make it possible to understand the significance and consequences of automated decisions; to obtain an explanation of an automated decision; and have the right to challenge the decision. Concerns over the interaction between workers and robots and in particular the presence of 'cobots' is the focus of the fifth dimension. Continuous feedback from workers on working practices as well as on the quality and volume of work, and a clear distinction between workspaces occupied by humans and workspaces occupied by robots, are the main propositions put forward by the ETUI.

Similarly, the ETUI's sixth dimension wishes to establish a hierarchy where human workers could and should be in control over robots and in turn the former should always be the priority of any employer. The seventh and final dimension stresses the need for all workers to become AI literate. This means learning to work alongside robots, AI and automation technological equipment and critically understanding the role of AI in the workspace. Schools and other educational institutions can play a vital role in assessing the current and future state of employment and in providing education for up to date skills and tackling employment uncertainty.

The 2020 budget of the newly elected Conservative government took place against the backdrop of the global outbreak of COVID-19. The Chancellor of the Exchequer Rishi Sunak (HM Treasury, 2020) emphasised that the fundamentals of the UK economy are strong, and the government is well prepared to protect public health and support economic and professional security throughout a temporary period of economic disruption. The budget was largely dominated by the government's 'levelling up' agenda of raising productivity and growth in all nations and regions and addressing disparities in economic and social outcomes (HM Treasury, 2020). But how far is the 'levelling up' agenda informed by the discourse of automation? In the first instance, automation and emerging technologies are conspicuously absent from the government's budget. The word technology in its multiple and abstract meanings features in 13 pages of a 121-page document. The aspirations of preceding governments regarding the effects of automation on growth, productivity, employment and wellbeing have given their place to capital spending for boosting science, technology, engineering and maths teaching. The government acknowledged that the UK's success in the global economy depends on innovation and technology and the budget will create the 'high quality', 'highly paid' jobs of the future without necessarily specifying how these new jobs will correspond to the increasing use of robotics, automation and algorithms.

With the rapid spread of COVID-19 and its effects on the economy and society, the government's otherwise ambitious budget was quickly side-lined by a series of emergency measures aspiring to guarantee wages, employment and the overall stability of the economy. When and where it is possible the economy needs to keep going; when and where it is not possible the economy needs to be on hold in order to be reactivated at a later point. One of the most striking features of the government's COVID-19 strategy is the restriction of individual movement according to specific types of employment. The government's unofficial classification of workers into 'key', 'essential' and 'frontline' workers indicates those whose work is essential to fighting the pandemic. Health and social care, education and childcare, local and national government, food and other necessary goods, public safety and national security, transport, utilities, communication and financial services constitute the new vital sites for maintaining social ties and supporting the economy.

As soon as the discourse on the economy shifts from growth and productivity to support and survival there is a parallel reconsideration of priorities and hierarchies. The prominence of capital and technology gradually wanes in favour of manual labour and social work, and the 'foundational economy' (Foundational Economy Collective, 2018) of housing, education, welfare and care becomes the first line of defence against the pandemic. By imposing emergency rules on when, where and how to work, the government implicitly or explicitly acknowledges the limitations of the labour market in securing jobs and protecting society.

The labour market and society: governing the (working) population

The current debate on automation highlights the dangers of reversing a recent yet very effective historical trend towards ever higher employment figures – a trend sustained by low wages, immigration and by the tendency of technological developments to create more jobs than they render obsolete. This overview of a series of reports, studies, budgets and public statements on the current state and future potential of automation indicates a specific view of the value of work as well as of the relationship between economy and society. The ideas presented and developed in the texts discussed aim to present the economy and the labour market as an apolitical field, governed by the otherwise unquestionable objectives of economic growth, productivity and competitiveness. These are objectives that unite think tanks, political parties and businesses regardless of their political inclinations. This consensus is supported by a tripartite relationship between national and global notions of the economy, the promise of automation to deliver economic growth and productivity, and concerns over the welfare and precarious status of workers.

The economy is at once national and global, caught in a process of constant territorialisation and deterritorialisation. Economic events and phenomena are territorialised within a national space and effectively have to be governed by the values, laws and policies emanating from the political authority of that space. The status, size and overall performance of the territorialised national economy becomes evident by the comparison and competition between national economies. A common theme in the texts is a series of comparisons concerning the current status of automation and its effect on productivity, growth, employment and employability skills. From these comparisons a clear sense of national economic strategy and policy emerges. Which countries aspire to a fully automated economy? What are the national responses to the imminent danger of mass unemployment? How are national education curricula responding and should respond to the advent of automation and robotics? Which country has the highest productivity rate? Which country has the fastest growing economy? What rights do workers have in each country?

However, the inevitable comparisons and competition between national economies do not only allude to the existence of a global economy but also to the rather limited capacity of the state to be the sole governing power of a national economy. As Nikolas Rose (1993: 339) points out, while governments still have to manage the national population, the growth, productivity and overall economic performance of a nation 'can no longer be so easily mapped onto another'. Inevitably, as the texts so clearly demonstrate, the national population can only be governed as long as it is fragmented and categorised according to region of residence, employment, education, income, age, gender and ethnicity. Yet, such a fragmentation does not necessarily lead to an effective democratic governance of the population. The social, economic and demographic fragmentation discussed in the reports cannot keep up with demands for recognition and participation in the economy. Would it be possible for national governments operating within the framework of capitalism and of the competitive market to address the numerous ways in which automation of work affects or might affect immigrant workers, women, older men in need of further training and young people entering the labour market?

Against the background of the governing authority and administrative limits of the state over the economy, there has been a narrow focus on policies and strategies aiming at increasing growth and productivity. In particular, what has become known as 'knowledge intensive business (KIBS)' and 'high end manufacturing' (European Monitoring Centre on Change, 2005; Foundational Economy Collective, 2018) dominate national and global responses to stagnant growth, low productivity and ageing demographics. Corporations engaging in AI, automation and robotics are seen as the agents of innovation, efficiency and productivity and most importantly of economic renewal and competition. In other words, a specific segment of the economy's private sector is perceived as the quintessential wealth creators, and at the same time trade unions, educational institutions and the welfare state have to play a supportive and proactive role.

The expansion of market relations and the creation of a competitive labour market can be detrimental to the social fabric. As early as 1944, Karl Polanyi argued that 'a self-regulating market demands nothing less than the institutional separation of society into an economic and political sphere' (Polanyi, 1944: 71). Polanyi (1944) understood that such demands tend to be utopian and outright disastrous. The commodification of labour and the subsequent establishment of a free, competitive labour market cannot exist for an extended period without compromising public infrastructure and most importantly human life. Social institutions such as education and welfare provide an essential 'protective covering' to human beings from exposing themselves to the dangers of the free, competitive labour market. Consequently, Polanyi (1944) argues that it is the state's role to regulate markets and by association to protect citizens through the strengthening of established institutions and the development of new ones. In Polanyi, we find a false yet persistent dichotomy between free markets and state intervention.

More specifically, contemporary left ideas about job security, workers' rights and education are based on the premise that corporations want to operate in an unrestricted economic environment whereas unions and progressive think tanks wish to reassert a more dynamic and interventionist role for the state. The relationship between the economy and the state is one of contradictions and demands. While all the reports, studies and public statements on automation are in agreement with Polanyi's argument for an interventionist state, there is a strong indication that the terms of intervention should prioritise the market over the state and reconfigure their relationship in order for the latter to be at the service of the former. As Dardot and Laval (2013) note, the state has been responding to free markets and transforming itself 'from within and from without'. From without, the ongoing privatisation of public services has severely limited the capacity of the state to be a producer of jobs as well as a protector of jobs and rights. From within, the state is subject to constant performance evaluations concerning the governance of the working population and its interventions in the competitive labour market. The maintenance and support of a competitive labour market becomes the priority precisely because it is the terrain in which all policies, predictions, political actions and interventions can be expressed and applied. It does not necessarily matter if the market is efficient, fair and adheres to competition rules, or if it is dysfunctional, unproductive and discriminates against certain segments of the working population. In any case, the competitive market must be protected as the mechanism for achieving the social and economic objectives of productivity, growth, prosperity, fairness and continuous education.

The understanding of the impact of automation as a temporary problem that eventually would be solved by the competitive labour market has been very popular among policymakers. The issue at stake is when an equilibrium will be reached where more jobs are created than disappear, and of course how this equilibrium will not challenge the current political and economic order. As early as 1927, the US Secretary for Labour James J. Davis expressed the need for a long-term perspective and belief in a system capable of self-regulation:

If you take the long view, there is nothing in sight to give us grave concern. I am no more concerned over the men once needed to blow bottles than I am over the seamstresses that we once were afraid would starve when the sewing machine came in. ... In the end, every device that lightens human toil and increases production is a boom to humanity. It is only the period of adjustment, when machines turn workers out of their old jobs into new ones, that we must learn to handle them so as to reduce distress to the minimum. ... We must ever go on, fearlessly scrapping old methods and old machines as fast as we find them obsolete. (Quoted in Frey, 2019: 175)

The case for uninterrupted technological progress at the service of the economy rests on the efficient management of 'the period of adjustment'. Policymakers, think tanks and unions fail to notice that the so-called 'period of adjustment' is a permanent state of affairs that serves the purpose of governing the working population by curtailing their demands and expectations. Fears of unemployment and of an unproductive workforce do not only reconfigure the relationship between the state and the economy but also legitimise a specific kind of government dealing with retraining, provision of skills and ultimately with economic growth and labour productivity. Such legitimation derives from the dual process of massifying and individualising the workforce. The immediate threat of mass unemployment needs to be evaluated in terms of the competitiveness of the labour force and its multiple fragmentations. With that in mind, all reports, policies and budgets acknowledge that some jobs will be incompatible with the future direction of the economy and eventually will disappear while new ones will emerge. At the same time, workers across all sectors of the economy have to act and be governed like individuals.

Notwithstanding demands for the protection and re-education of workers there is a shift from the integration of workers as a collective into a new social and economic order to a 'project of self-realisation' (Boltanski and Chiapello, 2007: 217) by establishing links between performance, knowledge and skills with the ability or even desire to remain employable in a competitive labour market. As a result, the flexibility that all reports implicitly or explicitly demand through reskilling and retraining transfer the uncertainty of unemployment and insecure employment from the corporation to the actual workforce. The new technological displacement is presented in the reports as an unstoppable, irreversible force whose mission is first to tackle low productivity and second to categorise the current and existing workforce according to the binary opposites of competent and incompetent, trained and untrained, skilled and unskilled, educated and uneducated, flexible and inflexible, male and female, young and old. The overarching emphasis on training and future investment also indicates the selection process of corporations regarding their future workforce.

The very presence of the 'cobot' as a technological artefact that embodies a phased process of automating the economy and a negotiated position between automation and unemployment becomes the perfect tool for assessing workers' performance. The cobot is generally defined as a smaller collaborative machine that is freeing robots from their confined spaces and one-dimensional tasks and allows human workers to collaborate with it. However, the cobot will not only minimise interaction between human workers but will also compete with human workers and indicate the latter's deficiency and need for training and improvement. Access to training will be offered to those 'whose disposition is deemed sufficiently promising to justify the investment' (Boltanski and Chiapello, 2007: 237). Once retrained and apparently employable, workers need to fulfil the promise of an automated, productive and growing economy, and will be indebted to the corporation for maintaining their employment status.

5

The Men Machines: Migrants as Robots

Who wants migrants?

What is the purpose of migration policy? How do governments decide who to include and who to exclude from their jurisdictions? Could a government ever be truly pro-immigration? Via the adoption of a genealogical approach, this chapter addresses these questions and puts forward the argument that the persistent crude cost–benefit logic that has underpinned migration policy throughout the 20th century, and that continues to inform contemporary migration policy, effectively reduces migrant workers to machines, that is, to non-human units of labour. The chapter argues that migrant workers could be understood as robots in the sense that they tend to be defined by their economic value and utility while also being denied their humanity. The argument is somewhat Aristotelian in that it is predicated on Aristotle's claim that a person who is capable of belonging to another and has to work for a living is to be understood as a slave (Aristotle and Saunders, 1995). In other words, this category includes those whose work is the use of their bodies, and even if this work is essential for the proper functioning of economy and society, it is not necessarily human *per se* (Aristotle and Saunders, 1995; see also Agamben, 2015).

The inescapable political and policy fixation on the importance of labour and of being economically productive, as well as the last liberal defence of immigration – the notion that migrants come here to work, and they do jobs that we don't want to do – is why this chapter focuses exclusively on economic migration.

Most people migrate because they are/want to be productive. Indeed, empirical evidence suggests that work and study are the key drivers of migration to the UK (Sumption and Vargas-Silva, 2020). Office for National Statistics (ONS) data (2020) also indicated that between 2010 and 2019, 40 per cent of migrants who moved to the UK for at least a year declared that the main reason for their move was work. In a recent report on work visas and migrant workers, Sumption (2021) has argued that an estimated 55 per cent of migrant workers who said that they had originally moved to the UK for work-related reasons were born in EU countries. In other words, from the mid-2000s up until the end of the Brexit transition period in December 2020, intra-EU mobility was the main source of work-related (specifically low-skilled and/or low-waged) migration to the UK.

Othering and bordering

State sovereignty and the ability of a state to choose who to admit and who to deny entry is seen as a prerogative in international politics and law (Nafziger, 1983; Abizadeh, 2008; Miller, 2010). And while arguments have been made that sometimes states admit migrants even if they don't really want to due to international legal constraints (Joppke, 1998), this is traditionally the case with asylum migration and other forms of mobility under the guise of international protection and humanitarianism. Things are slightly different with economic migration, which tends to be wanted (in varying measures) and codified (to varying degrees). The key takeaway point here is that economic migration policies could be seen as a proxy for or as an extension of fiscal policies; as an auxiliary policy tool to regulate the labour market, its racial hierarchies and conflicts, and to maintain low inflation rates due to the political ability to demonise migrant workers in public discourse, as well as to take away their labour and human rights. Thus, such policies are imbued with performativity as they do not simply regulate and manage work and workers, they actively contribute towards their creation.

The argument is predicated on a layered understanding of otherness and bordering. First, historically and politically the way migrants have been constructed as others, that is, as *xenoi* has become increasingly complex. The chapter traces this discursive othering from the early 20th century to the current post-Brexit context via a critical reading of immigration laws, policies and reports. While explicit preferences for migrants belonging to a particular race and ethnicity have become more covert over time, such notions remain firmly embedded in institutional processes and practices of immigration control, also amplifying the importance of social class and cultural capital. This has become evident in the aftermath of the 2008 global economic crisis and the subsequent climate of austerity and appears cemented in the post-Brexit context, indicating a nuanced and ambiguous discourse of othering, which simultaneously dehumanises migrant workers and justifies and legitimises inequalities.

Second, and related, economic migrants are particularly vulnerable to another layer of othering and bordering on the institutional level. This includes access to public services and provisions (for example schools, healthcare, social benefits, legal aid). Politically motivated decisions to limit such entitlements, or to make them conditional, regardless of evidence pointing out that migrants are not only less likely to use publicly funded services and benefits but also that not being able to access these puts migrants at a greater risk of exploitation, are analysed with particular reference to political and policy developments after the 2010 general election, up to and including post-Brexit migration policies and proposals. Public and political discussions and policy changes point to the continuous efforts of the political establishment to dehumanise migrant workers by implying that their very humanity, as opposed to their economic value, is what makes them fundamentally undesirable and problematic. This institutional and discursive othering and bordering is particularly important because of its securitised context and the very idea that anxieties about immigration, migrants stealing jobs and/or abusing the British welfare system are not only commonsensical, but also self-referential and self-validating. The argument has important theoretical implications for the sub-categorisation of Homo Oeconomicus along ethnic lines because low-skilled/ low-waged migration has been and remains racialised.

British immigration policy and the discursive production of *xenoi*: 1900s to mid-1990s

While it is beyond the scope of this chapter to provide a comprehensive overview of British migration policy, it seeks to demonstrate that it is marked by both continuity and change as concepts, such as social class, ethnicity and gender, are omnipresent and ever intersected, as well as transformed in political and policy discourses. The discursive production of others, that is, *xenoi*, could be explained through the prism of racialisation – a practice that could be either overt or covert, that could focus on physical appearance and/ or a set of behavioural characteristics but is quintessentially stigmatising and dehumanising to those deemed to be others (Murji and Solomos, 2005; Fox et al, 2012). Given that, for the most part of 20th century, British migration politics and policies were exclusively focused on race and ethnicity, it would be helpful to explore how these concepts played out in political, policy and legal discourses.

Prior to the 20th century, there wasn't a comprehensive body of laws, policies and regulations that were to control inwards mobility; rather, there were (sporadic) provisions that aimed at controlling the movement of aliens, that is, those who were not British subjects (Clayton and Firth, 2018: 2). The mobility of aliens tended to emerge on the agenda within the contexts of war, natural disasters and pandemics (Nafziger, 1983). The beginning of immigration controls the way we understand and experience them today could be traced back to the late 19th century and the widespread persecution of Eastern European Jews (Nafziger, 1983). Clayton and Firth (2018: 2) note that political, and public discussions were polarised as to whether Jewish migrants make a worthwhile contribution to British society or are driving down wages and fuelling low-level crime. Tony Kushner (in Murji and Solomos, 2005: 207–227) offers a critical discussion on the ways Eastern European Jews (but also Roma people) were racialised in Britain during the 20th century. He asserts that in 'the late 1940s and early 1950s, those from Italy and Eastern Europe were regarded as poorer alternatives to more Westernized immigrants such as those from the Baltic states' (Murji and Solomos, 2005: 221), claiming that their poverty was a key reason for the British to deem them undesirable.

Here, it is worth noting that there are other well documented historical instances of racialising phenotypically White migrants (McDowell, 2009; Fox et al, 2012). During the late 19th and early 20th centuries, there were attempts at racialising the Irish. The desire to control potential immigration 'floods' also appears to be well established, alongside concerns about increased criminality, anti-social behaviour and undesired labour market competition (Engels, 1845). Anglo-Irish differences were articulated in cultural rather than racial terms, and were a product of a specific historical moment, 'answering a need for psychic reassurance at a time when popular confidence in Britain's national mission had been shaken' (Douglas, 2002: 56, 57). The case of the Italian migrants in the 20th century is another telling example. In large US cities Italians were often associated with ethnic 'ghettoisation' (Glazer and Moynihan 1970: 186), limited if any knowledge of the English language, cheap labour undercutting the market (Glazer and Moynihan 1970: 192), being prone to crime and far too keen on their 'village' culture (Glazer and Moynihan 1970: 189–190). Such examples challenge the sometimes-assumed monolithic nature of Whiteness, shedding light on its fluid, contingent and contested nature, which produces fragmented degrees of belonging (Garner, 2006: 270).

Going back to the issue at hand, part of the government's response to public anxieties over the settlement and behaviour of Eastern European Jews in Britain was to set up a Royal Commission to investigate these concerns and the overall impact of Jewish migration on housing and employment opportunities, on public health and order. The findings of the report didn't indicate that these aliens were depriving the local populations of opportunities or that they were spreading diseases or being involved in criminal and anti-social behaviour. The Royal Commission nonetheless recommended the establishment of some form of border control, which resulted in the 1905 Aliens Act – the first piece of modern immigration legislation (Clayton and Firth, 2018: 3). What is noteworthy here is that the Royal Commission sought to provide a solution to a 'problem' that wasn't a problem in the first place, thus paving the path for its contemporary successors – the Migration Advisory Committee (MAC) being a case in point – to effectively disregard their own empirical research findings and expertise in order to provide policy solutions to issues that do not necessarily need them.

The Aliens Restriction Act came into force in 1914, during the time of the First World War (1914–1918), which granted extended powers to the Secretary of State (Clayton and Firth, 2018: 5, 6). Clayton and Firth argue that the implications of the 1914 Act have been far-reaching because these extended powers continued to apply during peacetime, despite the fact that they were introduced as an ad hoc measure during the war; and because the Act formed the legal basis of immigration control. The First World War period was also when visas were introduced, again, at first as an ad hoc measure but later on remained as a policy tool for immigration control if/ when an influx of people was expected – either from the Commonwealth, or from continental Europe (Clayton and Firth, 2018).

The Second World War (1939–1945) erupted at a time when there was a growing sense of (trans)national unity and belonging within the context of the British empire. Commonwealth citizens were praised – including by the British media – for their valuable military and non-military contributions to the war effort, both in the UK and across the Commonwealth nations (Clayton and Firth, 2018: 6, 7). It is worth noting, however, that in the period after the war as Britain was slowly but steadily moving towards a more multicultural society, there were some public and political anxieties about the increased migration to the UK from the Commonwealth (Hansen, 2000). In fact, up until the 1970s, official polling data suggested that not only was there significant public opposition towards Commonwealth migrants, but that this unease had been present on a cross-party level with both Labour and Conservative governments being averse to it (Hansen, 2000: 4, 5). Nonetheless, the period between 1948 and 1962 is regarded by Hansen as the laissez-faire years due to the significant volume of Commonwealth migration and the gradual societal transformation towards multiculturalism. Hansen discusses the issue of Commonwealth migration with a particular reference to the 1948 British Nationality Act, which via the trope of empire enabled substantial primary immigration towards Britain. In his words, 'the story of post-war migration is the story of citizenship, which was defined in the United Kingdom for the first time in 1948' (Hansen, 2000: 35). The 1948 British Nationality Act gave full citizenship rights to Britons and colonial subjects alike, thus creating facilitating conditions for increased migration from the Commonwealth to the United Kingdom (Hansen, 2000).

While some interest groups and anti-racist and pro-migrant civil organisations sought to challenge existing prejudice, everyday and institutional discrimination against non-White settlers, the Trade Union Congress' (TUC) position was more ambiguous and cautious. The inherent friction driving this position was between the anti-racism ethos of the TUC and its agenda to press for full employment and improved workers' rights, which in the TUC's view warranted some immigration control (Hansen, 2000: 7). Interestingly, it has been argued that the initial response of British trade unions to the Eastern EU enlargement has also fluctuated between rhetorical expressions of solidarity and anxiety about social dumping (Fitzgerald and Hardy 2010).

In the aftermath of the Second World War (1939–1945), Europe in general, and the Allies in particular, were faced with a substantial humanitarian problem: millions of displaced Soviet citizens, not all of whom could (or wanted to) be repatriated (McDowell, 2009: 21). Motivated by its economic needs more so than by humanitarian concerns, the then British government decided to recruit workers from the camps. The main reason was acute labour shortages caused primarily by the war, but also by the emigration of more than 1.5 million Britons towards Australia, New Zealand and Canada immediately after the war. Hundreds of thousands of workers – preference was given to people from the Baltics and Ukraine due to their phenotypical Whiteness – were recruited during the late 1940s to work in industry, mining and agriculture (men) and in hospitals, tuberculosis sanatoria and upper-class private households (women) (McDowell, 2009: 24–26; see also Fox et al, 2012). Government preferences were racialised and gendered: young, single and White women, especially from the Baltics, were recruited as well as praised for their appearance, personal hygiene and cleanliness and their potential to blend into British society due to their phenotypical Whiteness and cultural proximity. These women were thus to be differentiated from Caribbean female workers who were recruited around the same time. Interestingly, the TUC was vocally opposed to the post-war recruitment under the European Volunteer Workers (EVWs) scheme, undertaken by a Labour government, nonetheless. By way of an appeasement, the first round of recruitment was for single, female workers as the TUC was fearful of competition from male labourers (Hansen, 2000; McDowell, 2009).

Crucially however, the EVWs were effectively unfree labourers, who were regarded as a commodity and only valued insofar as they were fit and non-disabled (McDowell, 2009: 24). They had no rights and protections as part of their employment. Their legal status was both ambiguous and precarious: technically speaking, they were displaced persons who signed up (that is, volunteered) to participate in the British EVWs scheme. This meant that these people were neither refugees, which would have provided some humanitarian protection, nor workers in that they weren't waged, they were tied to their designated employer and so they couldn't change their employment, while also being denied basic working rights and/or the right to settle in Britain. It is worth noting that the Cold War era was marked by the notion that Britain had neither duty of care, nor owed any leniency towards people who were conquered by a rival power (in this case, the Soviet Union) (Glenny, 1993; Kopstein, 2009).

Thus, McDowell (2009) argues, while there are obvious differences between the 1940s EVWs and the influx of economic migrants after the 2004 Eastern EU enlargement, there are two important similarities. First, a racialised preference for White workers began to crystallise in the post-war period and it gradually became embedded in policy choices that continue to act as immigration filters today. Second, this Whiteness – understood against the backdrop of endemic, structural racism towards non-White migrants – is a form of privilege, even at the bottom end of the labour market, as it increases the opportunity for social mobility, while decreasing the risk of being (racially) harassed. However, and related, a key difference between old/new European economic migrants and post-colonial economic migrants is the lack of previous connections to the UK: most European migrants don't have an emotional attachment, family connections or any sense of shared identity with the UK. Also, unlike Commonwealth migrants, Europeans speak English as a second language and more often than not enter the British labour market with little if any understanding of it (McDowell, 2009).

While post-Second World War European migrants were unpaid and unfree, if racially desirable, the 1950s saw Commonwealth migration becoming increasingly racialised. Hansen (2000: 10) argues that successive British governments began to define this migration through the prism of colour in order to problematise it and, consequently, to push for more restrictive migration policies as a solution to a top-down constructed issue. Carter et al (1987: 135–157) argued that:

> The British state's policy involved direct intervention on some issues and an apparent inactivity on others. ... This went far beyond the prejudiced attitudes of individuals, albeit individuals holding high office. It amounted to the construction of an ideological framework in which Black people were seen to be threatening, alien and unassimilable and to the development of policies to discourage and control Black immigration.

The deliberate development of such a racialised framework meant that social issues related to employment opportunities, housing, crime and public order became intrinsically and causally linked to immigration, despite the lack of a clear link between these phenomena (Hansen, 2000). Thus, the post-1962 migration climate was marked by negative public and political attitudes, and further restrictive policies. Hansen (2000: 20) argues that 'once the door to the Commonwealth was finally closed, the closure was quick and complete. The rapidity alone is remarkable'. Both Conservative and Labour governments operated under the assumption that harmonious societal relations rested upon the implementation of ever stricter immigration control policies, which ended the privileged migration status to most Commonwealth migrants and postulated that family reunification is not a right that could be taken for granted (Hansen, 2000).

From the 1960s onwards, British immigration legislation expanded rapidly and quickly established itself as one of the strictest in Europe. Here, Hansen (2000: 26, 27) notes the role of the British executive branch of government which has been overly dominant and not subjected to enough scrutiny from the legislature. Following the UK's accession to the EU in 1973, some sense of Europeanisation of British public policy could be observed in that there was another, supranational level added to institutional processes and practices, but this never amounted to any actual political or policy restraints (Hansen, 2000; see also Flynn, 2005). A 1973 European Commission of Human Rights decision against the exclusion of Kenyan Asians from the UK was de facto ignored by the British government (Hansen, 2000: 28), setting the course for further opt-outs and acts of British exceptionalism down the road.

The legislative approaches of the 1960s and 1970s aimed to 'restrict severely the numbers coming to live permanently or to work in the United Kingdom' (Immigration and Nationality Directorate, 1996). Ever since the 1960s when temporary and guest workers in UK and elsewhere in Western Europe proved to be less temporary and less guests than originally envisioned in policy, and especially since the oil crisis of 1973, public and political attitudes against migration were on the rise. In this respect, the British policy swing from liberalism to restrictionism did not exist in isolation from its wider European and Global North context. In any case, with the oil crisis and the retrenchment of the old Fordist, mass employment industries, demands for economic migration began to decrease. This change was accelerated and exacerbated by the neoliberal ideology that came to dominate the 1980s and which resulted in ever greater levels of deregulation of the British economy (Flynn, 2005: 468).

Britain's EU membership and globalisation pressures resulted in frictions between the global and the national levels and demonstrated the limitations of state capacity within the realm of border control (Flynn, 2005: 465). Specifically, the increased volume of international travel from the 1980s onwards proved to be a policy challenge and was perceived as a crisis. Flynn (2005: 466) argues that under the provisions of the 1971 Act a relatively small fraction of people were subject to immigration control within the wider context of people passing through the UK and that up until the 1980s the assumption was that all travellers could be intercepted and examined at the point of crossing a physical frontier. Technological and communication advancements associated with globalisation rendered such assumptions moot, or at least unviable (Flynn, 2005; see also Hylland

Eriksen, 2007). Racialised assumptions of desirability contributed towards the increased reliance on pre-entry visa controls and screenings against people from some nationalities (Fitzgerald, 2019). For example, in 1985 nationals of Sri Lanka, Bangladesh, India, Pakistan, Ghana and Nigeria became the first Commonwealth countries to be entered on the British visa list (Flynn, 2005: 467).

The UK entered the 1990s with an ever so restrictive immigration policy that remained preoccupied with race, steeped in neoliberalism and with limited considerations for workers' rights. It was becoming clear that this approach was not fit for purpose and that it was not able to offer tangible benefits to British economic interests, or the wider British public for that matter.

The next sections of this chapter will explore in further detail the New Labour era and its migration legacy, as well as the post-2010 Conservative-led and dominated governments and their embrace of a securitised approach to migration. This is important because the road to Brexit was marked by the continuous interplay between New Labour's rational and technocratic detour from racialised immigration policy, and successive Conservative governments' insistence on mystifying and demonising economic migrants. In so doing, the chapter demonstrates that New Labour's technocratic and the Conservatives' mythical views on migration are similar inasmuch as they are different. The similarity lies precisely in the underlying logic that migrant workers are not quite humans – rather, they are machines that should always be productive, achieving ever greater economic growth, and not demanding much of their hosts and operators. By highlighting the needs and niches of the British economy that require filling, migration policies have the potential to de facto make migrant workers more desirable than the British-born workforce because they can legitimise and officially sanction unequal and/ or discriminatory practices.

New Labour's managed migration policy: a paradigm shift?

The migration legacy of the New Labour era was marked by a technocratic approach to economic migration within the context of an increasingly Europeanised migration policy. The Eastern enlargement of the EU accelerated the multitude of political and economic interdependencies between the Eastern and Western halves of Europe and resulted in a substantial increase of the number of EU nationals coming to the UK (Arcarazo and Wiesbrock, 2015, Sumption and Walsh, 2022). New Labour governments were determined to detoxify migration debates by focusing on the economic utility of migration, which was somewhat facilitated by the then booming British economy and somewhat hindered by growing

Euroscepticism and a "stubbornly non-European British self-identity" (Dennison and Geddes, 2018: 1150).

The logic that labour migration is to be inextricably linked to the needs of the economy was pioneered by New Labour but remained a cornerstone for migration policy for later governments as well. This logic has been deeply embedded within political thinking and has effectively reduced any liberal defence of immigration to its economic utility, the latter being seen as the least objectionable, common denominator that everyone should be able to agree on. Indeed, the examined empirical materials reveal that migrants' ability to enter the British jurisdiction and their immigration status are exclusively defined by their ability to be economically productive. In other words, the work ethic, productivity and endurance of migrant workers mirrors the way people talk about robots while the human side of the workers (reproduction, access to social housing and benefits) is seen as socially and politically unacceptable.

Here, we focus more narrowly on EU workers due to the following reasons. First, Western Europeans' own sense of cultural and economic superiority over Eastern Europeans framed the Eastern enlargement as a civilisational quest (Mälksoo, (2006: 288), Buchowski, 2006) and in so doing cemented racialised and fragmented understandings of Europeanness. The Eastern EU enlargement has been a lengthy process, marked by asymmetric power relations as the very institution of candidacy became symptomatic of the simultaneously running processes of recognising Eastern European states' renewed claims to Europeanness and the contestation of these claims by emphasising that the candidate states were substantially lagging behind the core member states (Moravcsik and Vachudova, 2003; Rumelili, 2004; Asmus, 2008). These dynamics have contributed towards differentiated policy treatment of Eastern EU nationals who have been subjected to up to seven years of transitional arrangements. These arrangements, in conjunction with existing negative stereotypes towards Eastern Europeans, have jeopardised their working rights and access to welfare and public provisions (Pencheva, 2020b, Ulceluse and Bender, 2022).

Second, following from the first round of Eastern EU enlargement in the early 2000s, there was a lot of economic optimism in the UK regarding the possibility to rely on a willing, culturally proximate and legally straightforward pan-European pool of workers. Traditional migration costs (for example sponsoring work visas, work permits and so on) that would normally be borne by employers and would involve a complex bureaucratic machinery were removed. More specifically, they were transferred onto individual workers (Ciupijus, 2011), which theoretically streamlined the labour market integration of European workers from accession states.

Third, a key objective of the Eastern EU enlargement was to regularise the cross-national labour mobility of Eastern Europeans, thus transforming yesterday's irregular migrants into EU citizens (Fassmann and Munz, 1994; Finotelli and Sciortino, 2013; Arcarazo and Martire, 2014). However, within the British context, following the three rounds of Eastern EU enlargement (2004, 2007 and 2013), Eastern Europeans from accession countries have gone full circle: from non-EU migrants to EU citizens only to be degraded back to non-EU migrants (along with all EU nationals from the older EU-15 member states) following the result of the 2016 referendum. This denigration of substantive citizenship rights has been communicated and enforced through restrictive migration policies that nonetheless continue to be focused on the economic utility of migrants and the best ways this could be harnessed for the needs of the British economy.

Lastly, while evidence has been unequivocal that EU workers are economically beneficial for the UK and that they contribute more than they use in public services (see here Dustmann and Frattini, 2014), political and public unease with the pace and scale of this mobility has demonstrated how politically problematic they are. The perception that intra-EU mobility equals an unrestricted and uncontrollable mass migration was a key tenet for the Leave campaign in 2016 (Pencheva and Maronitis, 2018).

New Labour's migration policy marked a distinct break from the toxic obsession with race relations in the years after the end of the Second World War (1939–1945), yet continued to be restrictive with regards to asylum migration. The main building block of New Labour's approach to migration was to wed migration and economics as a way of detoxifying migration debates. A technocratic economic approach to migration policy was seen as pragmatic and unobjectionable, thus, an acceptable compromise for political parties from across the political spectrum (Somerville, 2007). Indeed, during Labour's second term (2001–2007) the unequivocal value of properly managing economic migration was upheld in policy, legislation and political rhetoric (Somerville, 2007). An essential component of the New Labour migration agenda was the establishment of the independent, expert-led MAC, which was tasked to produce, commission and evaluate evidence on the economic and social impacts of migration and to provide expert and impartial advice.

As early as the year 2000, the Labour government carried out a review of the impact of international migration on the British economy and its competitiveness, or lack thereof, and made steps towards opening up the labour market to highly skilled and entrepreneurial migrants (Somerville, 2007: 29). It is also worth noting that apart from expanding and debureaucratising the work permit system, New Labour also introduced a points-based system (Somerville, 2007: 30). Overall, the party has sought to fully embrace the liberal consensus and globalisation. For instance, unregulated short-term labour mobility for the provision of services under the General Agreement on Trade in Services (GATS) treaty was endorsed by New Labour. The focus of the major overhaul of both migration policy and the administrators responsible for its implementation was the needs of the employers, not the employees. Such logic has contributed towards migration policy becoming a key element of broader fiscal policies. Conversely, the development of a points-based system was based on the marketable skills and earning potential of prospective migrants – the higher the formal qualifications and skills, the easier it was to obtain a work permit.

Low-skilled migration was managed according to sectors of the economy, such as hospitality and food-processing. The Sector Based Schemes (SBS) were created in 2003 and gradually phased out by 2010. There was also the Seasonal Agricultural Workers Scheme (Somerville, 2007; Rolfe et al, 2013). The decision of the then Labour government to allow unconditional access to the citizens of those countries that accessed the EU in 2004 was a key factor for the dismantling of specific policy routes for low-skilled migrants as these routes were already dominated by Eastern Europeans. The larger than expected influx of Poles and other Eastern Europeans meant that the 2007 accession of Bulgaria and Romania and the 2013 accession of Croatia were gradual and conditional, with preference for the highly skilled and the self-employed/self-sufficient.

In maintaining a hard stance on asylum migration versus adopting a more open and pragmatic approach towards work migration, New Labour's migration approach reaffirmed the existence of two categories of migrants: *good migrants* – those who are economically productive and who contribute to the public purse via taxation; and *bad migrants* – those who constitute a drain on the public purse and whose mobility is politically problematic, if not altogether undesirable.

In this respect, it is worth considering the creation of the MAC. A key legacy from the New Labour era, the MAC has seemingly transcended political boundaries and is still fully operational, having worked during the Coalition government and subsequent Conservative governments. Despite the key emphases on independence and impartiality, the work of the MAC has not been free of controversy, particularly regarding its unique relationship with the Home Office. The MAC is situated within the same building as the Home Office, makes use of the same facilities, and derives all its funding from the Home Office budget. Inquiries and reports are also exclusively commissioned by the Home Office, including the Home Secretary and the Minister for Immigration.

MAC was established in 2007 as a 'a non-statutory, non-time limited non-departmental public body established and funded by the Home Office. It is comprised of a Chair and four other committee members who are appointed as individuals to provide independent and evidence-based advice to the Government on migration issues' (Home Office, 2014). Its main role is to advise the government on matters of migration on the basis of evidence and impartiality. In other words, MAC is to help British governments maximise the utility of migration for the British economy. A triennial review of the MAC carried out by the Home Office between 2012 and 2014 revealed that public perceptions and debates around immigration are often too 'emotional' as opposed to grounded within the unquestionable realm of facts and data (Home Office, 2014: 4). If the government has all the facts and figures then it can make truly informed decisions, which will maximise the benefits for the British economy, while minimising the risks. The review unequivocally indicates that a technocratic, rational-choice type of reasoning is the preferred epistemological guiding principle for corporate governance. Evidence, and particularly numerical evidence, is seen as apolitical, objective and therefore the most suitable guidance for migration policy. As such, the MAC team is divided into two parts: (1) analysis and research; and (2) policy. The former team is exclusively composed of economists, and in the latter team there are different members of staff responsible for corporate partners engagement, finance, communication and evidence.

However, this is not to suggest that New Labour's alleged value-neutral view that work migration should be subjugated to the needs of the British economy was completely untainted by race and ethnicity. On the contrary, it was based on a racialised logic and ended up (re)producing racialised hierarchies of entitlement within the British labour market. New Labour's managed approach to migration contributed to the differentiation between the 2004 and the 2007 cohorts in policy terms (Somerville, 2007; Fox et al, 2012). The UK applied transitional arrangements for Bulgarian and Romanian nationals, which aimed at providing stricter regulations in terms of their access to the British labour market. These were reviewed periodically based on MAC advice and maintained for the full period of seven years (Somerville, 2007; Rolfe et al, 2013). MAC's recommendations were informed by the economic situation in the UK, as well as considerations about the positions of other member states on the freedom of movement of Bulgarians and Romanians. As members of the European Economic Area (EEA), Bulgarian and Romanian citizens did not need permission to enter and reside in Britain, which however was subject to them being able to financially support themselves and their dependants (Rolfe et al, 2013: 1, 2).

Then Home Secretary John Reid and the minister for immigration Liam Byrne were both quoted in the *Guardian* praising the MAC as 'independent' and 'made up of business and union leaders', who 'would take into account the economic, tax and wider social impact of migration' in order to 'generate a more open debate about the level of immigration that was good for Britain' (Travis, 2006: 11). A strictly utilitarian and rational approach to migration was also endorsed by then Shadow Home Office Minister Damian Green, who was quoted in a news article in the *Daily Mirror*. He praised the decision to block 'unskilled workers from Romania and Bulgaria from entering Britain when their countries join the European Union' by claiming that 'we should have roughly the same policy so that we take in only the people who will benefit our economy' (Prince, 2013: 26).

However, the then government drew an even sharper distinction between EU and non-EU migration. The Home Office (2006: 6) was adamant that:

> With an expanded European Union there is an accessible and mobile workforce already contributing to our growing economy, closing many gaps experienced by employers. In a changing environment where our European commitments provide many opportunities for the UK to benefit from this new source of labour. ... Our starting point is that employers should look first to recruit from the UK and the expanded EU before recruiting migrants from outside the EU.

Indeed, during the early and mid-2000s the atmosphere was one of economic optimism that EU workers are good for the British economy. In terms of politics and policy, intra-EU mobility was narrated and presented as less problematic in comparison with non-EU and/or asylum migration. Related, it is important to point out that UK policymakers did *not* communicate the labour mobility of Eastern Europeans as an exercise of EU citizenship. Rather, it was meant to fill vacancies at the low end of the British labour market: a move which some have interpreted as a way of transforming the exercise of EU Treaty rights into state-managed labour migration (McDowell, 2009; Ciupijus, 2011).

Migration policies after New Labour: from technocracy to securitisation

The 2010 general election produced a Coalition government between the Conservatives and the Liberal Democrats and marked a discursive shift from the technocratic and managerial approach of New Labour to a more securitised understanding of labour migration. The securitisation of migration refers to the deliberate political choice to elevate migration from the realm of habitual, or 'normal', policymaking to the realm of exceptionalism (Wæver et al, 1993; Buzan et al, 1998). This practice has its roots in the so-called widening of the security agenda (1970s and onwards) which incorporated a wider range of non-military threats into the field of security studies (Buzan et al, 1998). While human migration is not a new phenomenon per se, nor is its treatment as an issue of security concern (Terriff et al, 2000: 157), the 'newness' comes to suggest that these phenomena have been analytically neglected during the Cold War era due to a different dominant rationale of viewing threats and vulnerabilities (Terriff et al, 2000; Huysmans, 2006; Dannreuther, 2007; Aguis, 2010). Further, scholars highlight the contradiction that while 'the Cold War was fought, at least in part, for the principle of free movement of peoples, in reality, its end has led to a distinctively less benign environment for international migration' (Dannreuther, 2007: 104; see also Huysmans, 2006). In Europe, 'the prospect of mass East–West migrations was viewed as a source of insecurity rather than as a positive assertion of human freedom' (Dannreuther, 2007: 104). Such anxieties were aided by the large numbers of refugees following the ethnic conflicts in the former Yugoslavia, as well as the increased number of irregular workers from the former Warsaw Pact countries (Fassmann and Munz, 1994; Huysmans, 2006, Black et al, 2010).

The shift from technocracy to securitisation that ensued following the end of the New Labour era was not only discursive, but also empirical in its appreciation of the role of evidence for migration policymaking. While for the politicians and policymakers of the New Labour era empirical evidence was the key to demystifying and depoliticising labour migration, the securitised discourse of the 2010s and onwards had a less straightforward relationship with evidence and its impact on policymaking. In fostering a breakup with the politics of racialisation, which dominated most of the 20th century, the architects of New Labour – via the motto 'what matters is what works', privileged the importance of economic (read, numerical) evidence, which had a dual benefit. On the one hand, it aimed at detoxifying migration debates (Balch and Balabanova, 2011), while on the other hand it boosted the credibility of the government by the performativity of competence, expertise and epistemic authority (Boswell, 2009).

The followed shift towards a securitised understanding of migration was predicated on the assumption that (labour) migration is a highly contested issue and that the fairly successful labour market integration of EU workers, coupled with the fact that they contributed more to the public purse than they used, was no longer enough to mollify the public. Empirical research into the communication of evidence and expertise in media discourses, for example, has highlighted that the tension between knowability and unknowability leads to an 'aversion to complexity' in communication and a more selective and partisan approach to evidence (Balch and Balabanova, 2011: 902). Further Pencheva (2019) has pointed out that British print media displayed the tendency to ignore and/or override evidence in order to advance a securitised understanding of labour migration and that while newspapers mentioned evidence (government data and statistics, industry, think tank and academic reports, and so on), said evidence had little to no relevance to the news story/editorial piece. The notion that the political establishment, as well as the British public, was falling out of love with empirical evidence and expertise found an expression in Michael Gove's (2016) infamous statement in the runup to the 2016 referendum on the

UK's membership in the EU, that "people in this country have had enough of experts".

Although the differences between New Labour's technocratic, managerial approach and the post-2010 shift to a securitised approach are obvious, there is one important similarity: both approaches attempted to distance themselves from the explicit focus on race and ethnicity, thus seeking to avoid accusations of racism. In other words, while neither (New) Labour nor Conservative governments were pro-immigration, neither wanted to be perceived as racist.

The new Coalition government inherited the functional division between *good* and *bad* migrants produced by the policies of New Labour but deepened it further via pushing through austerity policies and implementing the so-called hostile environment. A key deliverable policy objective under the Coalition government was the reduction of net migration (the annual difference between the levels of immigration and emigration) from 'the hundreds of thousands to the tens of thousands', as well as limiting the rights of new and resident migrants to access social benefits and other public services. In order to meet this target, a set of policy measures was introduced, including a cap on skilled non-EU workers, minimum income requirements for family reunification with a non-EU family member, as well as stricter government guidelines on sponsoring non-EU students (Sumption and Vargas-Silva, 2020). The objective of reducing the levels of net migration was pursued by three consecutive governments following the 2010 general election, only to be abandoned by the 2019 Conservative government under Boris Johnson due to its unviability (Sumption and Vargas-Silva, 2020). A clear rationale, or indeed, any form of political and policy justification as to why net migration below 100,000 is good while levels above 100,000 are bad, was never provided (Robinson, 2013). The transition from the technocratic to the securitised was facilitated by the objective fact that there were more migrants coming into the UK (Hawkins, 2013), and the securitised media and political discourses of how socially and politically catastrophic that is.

The first round of Eastern EU enlargement in 2004 contributed towards a rise of net migration to 268,000 persons per annum, peaking at 331,000, reached in the year ending March 2015 (Sumption and Vargas-Silva, 2020). In other words, the significant influx of economically active people under the EU's freedom of movement principle – largely seen as a permissive migration regime – resulted in a sense of collective panic that quickly became self-referential and embedded within a hyperbolised and non-factual discourse. It is paradoxical and worth noting that despite the heightened anxieties around the mobility of EU citizens, the net migration targets were originally aimed at non-EU migrants whose mobility is more sternly regulated anyway.

This paradox is politically and analytically important, because up until the mid-2000s, both political parties were playing fast and loose with the term 'immigration' and their manifestos remained focused on controlling irregular and non-EU migration, as well as modernising the asylum system, rather than reducing the levels of intra-EU mobility. In fact, chapter 7 of the 2005 Labour manifesto stated that the party was proud to be a member of the EU and that 'outside the EU (Britain) would unquestionably be weaker and more vulnerable' (Labour Party Manifesto, 2005: 83). Labour promised, like the Conservatives (Conservative and Unionist Party 2005: 26), to hold a referendum on the membership of the European Union and it openly declared the position that the party would fight for a 'yes' vote in such a referendum. While both parties promised a referendum, neither were outwardly against European immigration; indeed, both were keen to expand the borders of the EU knowing the free movement of labour would still apply, even with restrictions.

The British media environment was a significant factor here because a complex and contested issue such as migration would inevitably attract media attention (Balch and Balabanova, 2011; Daddow, 2012). Empirical research suggests that between 2001 and 2016 migration was consistently listed as a top public concern for the British people (Blinder and Richards, 2020). In 1994, which was the starting point of this data series, less than 5 per cent of respondents thought of immigration as a concern, and it remained rarely mentioned prior to 2000. In the year or so before the EU referendum, between June 2015 and June 2016, immigration was consistently named as the most salient issue facing the country, peaking at 56 per cent in September 2015. A study commissioned by the United Nations High Commissioner for Refugees and carried out by the Cardiff School of Journalism explored press coverage of the refugee and migrant crisis in the EU across five European countries – Spain, Italy, Germany, the UK and Sweden. Researchers did a content analysis of articles published in 2014 and early 2015. The report concluded that, overall, the Swedish press was the most positive towards refugees and migrants, while coverage in the UK was the most negative and the most polarised. The researchers argued that Britain's right-wing media was particularly aggressive in its campaigns against both refugees and economic migrants.

Academic and journalistic research has revealed that migration-related coverage has remained overwhelmingly negative, and that positive material tends to be omitted. Emphasis on repetition, disinformation and practices of overriding empirical evidence have resulted in the negative prominence of migration in public discourse. For instance, a journalistic investigation by former *Sunday Times* columnist Liz Gerald found that the *Daily Express* had carried 179 front pages in five years devoted to anti-migrant stories, while the *Daily Mail* had published 122. The prominence of migration coverage, as well as the growing public concerns around the social costs of migration, have been amplified by the tendency of tabloids and broadsheets to oversimplify the complex issue that is migration, to cherry-pick and/ or ignore evidence, as well as to suggest that there is an easy (and obvious) solution in the form of increasingly selective and restrictive migration policies (Balch and Balabanova, 2011; Pencheva, 2019).

Blaming immigration for various social issues also became a prominent characteristic of political discourse with then Prime Minister David Cameron blaming New Labour's migration legacy for being too lax. Although leaders of both the Labour and the Conservative parties have advanced a robust defence of national sovereignty over welfare provisions, the Conservative-led Coalition government officials were more assertive in their conviction that EU institutions were dysfunctional and lacking legitimacy, therefore posing an existential threat to British interests (Daddow, 2015: 84). Key Conservative political figures, such as Iain Duncan Smith and David Cameron, have consistently referred to the British welfare system as too generous and fragile. Then Prime Minister David Cameron proposed welfare policies changes, which local authorities were to adopt in order to restrict the eligibility of EU nationals to claim benefits, and to be able to 'deport those found homeless' (Copsey and Haughton, 2014: 79). The larger than expected influx of workers from Eastern EU member states facilitated the securitisation of the British labour market and public services as attention began to shift from the economic utility of these migrants to the social costs of their mobility.

For instance, as the end of transitional measures for Bulgarian and Romanian nationals on 1 January 2014 was nearing, British mainstream media was increasingly preoccupied with the alleged risks unfettered access to the British labour market and welfare system would pose (Pencheva, 2019). A political commentary from the *Mail on Sunday* in December 2013 quoted a leaked Home Office report which warned of potential tensions between Bulgarians and Romanians and Poles over employment opportunities at the low-end of the labour market, as well as for housing and social services. The newspaper referred to an interview conducted with 53-year old Sharon McNeil from Maidstone, Kent – one of the constituencies in the southeast with the highest concentration of Eastern Europeans. Ms McNeil expressed concerns that "there's no room at the schools or extra places. The services are stretched. At my doctor's surgery you can't get an appointment for at least two weeks." After briefly referring to a report from the University of Reading which argued that the impact on public services will be limited in the short term as most Eastern Europeans are young, healthy and only rarely bring any dependents, the *Mail on Sunday* dismissed it and claimed instead that: 'there will be an impact on schools as one in three Bulgarian and Romanian immigrants are expected to have school-age children – and one in four will have babies – within four years of arriving' (Petre and Walters, 2013: np).

This is just one example of the concerns raised by British mainstream print media that while migrant labour could be beneficial under specific circumstances, migrant workers settling down, falling ill or getting pregnant is a different question altogether (Pencheva, 2019). In other words, discursive accusations that Eastern Europeans were stealing the employment opportunities of the locals, and that they put pressure on public services and provisions were gaining traction.

Specifically, the hostile environment was engulfing EU citizens by making it increasingly difficult to claim unemployment and other social benefits, as well as by reinforcing the transitional labour arrangements for Bulgarians and Romanians (2007 EU accession cohort) and Croatians (2013). Growing fears over the actual and potential impact of intra-EU mobility on national welfare systems have preceded every EU enlargement round and have been exacerbated by other external factors, such as economic recession and rising unemployment (Carrera et al, 2015). While earlier entrants, such as Greece, Spain and Portugal, have also been subjected to seven years of restrictions during a transitional period and there have been fluctuations in levels of mobility throughout and after this period, Kvist (2004: 307) argues that not much can be learned from history. One of the main reasons is that the migratory surplus of Eastern Europe was more substantial in comparison with those of Greece, Spain and Portugal. The combined Eastern enlargements of the EU in the 2000s have resulted in a relative population increase of the EU of about 27 per cent, compared with the combined Southern enlargements of 1981 (Greece) and 1986 (Spain and Portugal) (Kvist, 2004: 305). However, on average the GDP in the acceding countries is only a quarter of the EU-15 average, which is less than for all preceding enlargements, except the accession of Greece in 1981 (Kvist, 2004). Then Prime Minister David Cameron expressed a concern that some EU citizens might move to the UK under the freedom of movement rules but won't have the intention of getting a job.

A leaked government report (reported in *The Times*) suggested barring EU nationals' access to welfare payments for the first five years and limiting all labour movement from poorer countries who are new EU members until their GDP per capita is 75 per cent of Britain's. As a result, the government has raised the prospect of a 75,000 cap on annual EU immigration 'as part of a radical change in Britain's relationship with Europe'. The report acknowledges the existence of 'widespread public concern' about the so-called benefits tourism among political elites and the wider public alike. The article adds that the government is 'very concerned that the relative generosity of our in-work benefits acts as a strong pull factor for migrants and encourages benefits tourism'. Further, the text cites a polling by Lord Ashcroft (Conservatives), which found that '63 per cent believed immigrants were claiming benefits and public services when they had contributed nothing in return' (Leppard, 2013: 11).

Further, the Welfare Reform Act 2012 introduced tougher penalties for people who commit benefit fraud and postulated that benefits can be reduced or stopped if a claimant, their partner or a family member is convicted of a benefit fraud offence or accepts an administrative penalty for such an offence, that is, a financial penalty offered by the Department of Work and Pensions (DWP) as an alternative to prosecution (Welfare Reform Act, 2012). Under the Conservative-led Coalition government, the UK had sought to challenge the EU with regards to what it perceived as the unconditional eligibility for social benefits for EU nationals. David Cameron's (2013) Bloomberg speech in January 2013 promised an in-out referendum as a way of addressing the alleged democratic deficit between the British people and the EU. EU nationals' eligibility for various welfare payments, exacerbated by the impending dropping of labour market restrictions for Bulgarian and Romanian nationals, were high on the political agenda and prompted Cameron to tour EU capitals, seeking support for restricting the eligibility of EU nationals to access and export certain UK benefits (Copsey and Haughton, 2014; Carrera et al, 2015).

The anxiety around welfare dependency was exacerbated by the fear that mobile EU citizens from Eastern EU member states would be more interested in generating *income* in the broader sense of the word, as opposed to simply being in waged employment. Therefore, the main difference between the New Labour and the Coalition government's discourses is the decreasing focus on the economic utility of migrant workers in the latter. In other words, another key anxiety with regards to the welfare and public services theme is the growing realisation that workers also have human needs. Indeed, the very nature of 'social raids' is that getting work is used as an entry ticket into the national welfare system, 'sometimes with a view to exporting benefits to the home country' (Kvist, 2004: 313). These concerns share a belief that national labour markets and welfare systems become open to nationals from other countries, and that this may result in people shopping around to get the best mix of benefits, wages and taxes (Kvist, 2004). The risk of such social raids is exemplified by bogus self-employment, or by undercutting wages as a strategy to claim in-work benefits. Not only does the notion of economic utility fade after the New Labour era, its very essence has been fundamentally challenged by subsequent governments, while at the same time workers' rights have been trampled and the fallout of precarious and insecure work has been blamed on migrants themselves. However, it is important to note that the government had little actual control over intra-EU mobility, which was a particularly large migration stream into the UK with over 35 per cent of all immigration to the UK in the year ending September 2012 (Robinson, 2013: 5).

The significant inflow of EU workers from new member states and the inability of the British government to stop these people coming to the UK in response to rising public neurosis over the levels of intra-EU mobility were coupled with serious misunderstandings about the nature of the freedom of movement principle. The EU Directive on Free Movement (2004) has extended the right of free movement to non-gainfully employed (inactive) EU citizens, while at the same time this group of persons has been given access to the welfare benefits of host countries (Ochel and Sinn, 2003: 313). Moreover, the right of residence of gainfully employed EU citizens (employees and self-employed persons) has been broadened. So even though the Directive was not implemented in national laws and regulations until 2006, the notion that the freedom of movement will impose excessive demands on the solidarity of EU citizens in the host countries has been significant (Ochel and Sinn, 2013: 330). This has been framed as the EU basically giving the green light to any impoverished Eastern European to enter the UK, with or without a job, and access a generous welfare system, posing questions about the sustainability of national welfare states as the rounds of EU enlargement expand the pool of potential claimants.

Thus, the right of EU nationals to claim social security payments in the UK tends to be perceived as absolute and unfettered, even though it is by no means unqualified. Public anxieties and concerns were centred on the issue of migrants accessing the so-called 'special non-contributory benefits', thus sustaining the perception that some people would be motivated to exercise their Treaty rights solely to claim their entitlement to such 'minimum subsistence benefits accessible to the economically non-active' (Carrera et al, 2015: 259). Both media and political discourses opted to dismiss conclusive evidence (Guild, 2002; Carrera et al, 2015) that EU legislation contains a number of safeguards to prevent the abuse and misuse of the social rights provided and that national governments retain considerable sovereignty within the domain of social policy, as European social policy predominantly comprises market-enabling measures securing the free movement of workers (Pierson, 2001). Further, there is evidence that 'exogenous pressures stemming from European collaboration are far less important than endogenous pressures such as ageing populations' (Ochel and Sinn, 2003: 301; see also Esping-Andersen, 1999; Pierson, 2001).

While it is undeniable that differences exist between the capacity – and generosity – of welfare systems in the wealthier core member states, and in the poorer new member states, the actual significance of the so-called welfare magnets for (re)shaping migration flows is highly disputed (Borjas, 1999; Ochel and Sinn, 2003; Nowaczek, 2010: 290). Analysing data from the European Community Household Panel (ECHP), De Giorgi and Pellizzari (2009) demonstrate that migrants in the EU-15 states, that is, pre-Eastern enlargement, choose the destination countries they wish to move to on the basis, among other things, of the generosity of its welfare provisions (De Giorgi and Pellizzari, 2009: 354). Indeed, the underlying rationale of EU legislation is that if the social security position for EU citizens, in particular those who are workers, is not ensured in the host member state, EU citizens are discouraged from making use of their free movement rights in the first place (Carrera et al, 2015: 253). While the processes of economic and social integration have led to the harmonisation of different policies across the EU, very little has been done on the welfare side, maintaining significant heterogeneity of welfare provisions across different European countries. The authors argue that the issue is not so much whether migrants would rush into countries with generous welfare benefits, but to what extent the variation in the welfare institutions across the EU would generate distortions in migration flows (Carrera et al, 2015: 361). The 2008 financial crash and the subsequent Eurozone crisis have been significant as they have changed EU migration patterns. This is reflected in the data, which reports on the virtual halt of mobility from Eastern to Southern Europe and redirecting it instead towards Britain (Thompson, 2017: 439).

Although academic literature, as well as think tank reports, find little empirical evidence to support the assumption that welfare states with generous benefits and accessible labour markets will become magnets for welfare migration, it has been argued that following the Eastern enlargement, the core member states do enter 'in strategic interactions as if such migration would occur' (Kvist, 2004: 301). Such perceptions suggest that the British state is increasingly unable to control its borders and that its ability to use the national welfare system as a 'political filter' is also diminished (Geddes 2001: 21; Nowaczek, 2010).

Brexit and post-Brexit migration policies

From a relatively unproblematic policy issue prior to the 2004 EU enlargement, intra-EU mobility became a key argument for Brexit. The issue was deliberately side-lined by the Remain campaign where there were virtually no references to the economic benefits of it, while the Leave campaign overemphasised the social and economic downsides of intra-EU mobility. The dominant understanding from the New Labour era that migration policy is an auxiliary tool for fiscal policy and that EU workers should be celebrated for their productivity and high levels of employability had given way to the notion that EU migrants are posing unfair labour market competition to the local workforce by accepting to work for lower wages. In fact, the perceived threats of intra-EU labour mobility contributed towards the idea that the UK has no control over its borders and has no say in selecting migrants who are 'good' and 'useful' for the British economy. There was a growing realisation that British sovereignty is somehow an intrinsic part of British identity which has been abused by the undemocratic EU. The political communication of the Remain and Leave campaigns reflected this understanding (Pencheva and Maronitis, 2018).

Recent ONS data point out that net migration in 2020 for EU nationals was negative, with 94,000 more EU nationals estimated to have left the UK than to have arrived (ONS, 2020). While the global COVID-19 pandemic has certainly had an impact on this trend, the significance of Brexit as a factor should also be stressed. Despite solid empirical evidence that the levels of EU migration to the UK have been steadily declining since 2016 (Vargas-Silva and Walsh, 2020; ONS, 2021), EU workers continued to be treated like something that wasn't quite a problem but it had better get fixed anyway. The 2018 MAC report on the EEA migration in the UK is a good example of how the evidence-based approach of New Labour and the securitised narrative of successive governments can get discursively intertwined. The report was commissioned to evaluate the comprehensive impact of EEA migrants on different spheres of the UK economy, such as productivity, employment and housing prices, among others, in line with the government's Industrial Strategy, and was also tasked to make evidence-based recommendations on what a post-Brexit migration system should look like. The report essentially performs a cost–benefit analysis and concludes that:

> [D]espite the significant scale of migration from EU countries over the past 15 years, the overall economic impacts have been relatively small with the main effect being an increase in population. EEA migration as a whole has not harmed the existing resident population overall, as has been claimed by some, but also has not had the significant benefit claimed by others. (MAC, 2018: 110)

The overall tone of the report is neutral and detached as it seeks to portray EEA mobility in an objective, ideologically untainted manner. Although the report indicates that there are no significant negative impacts of EEA mobility to the UK, it asserts that: 'The problem with free movement is that it leaves migration to the UK solely up to migrants and UK residents have no control over the level and mix of migration. With free movement there can be no guarantee that migration is in the interests of UK residents' (MAC, 2018: 4, 5). The report puts forward a proposition that a post-Brexit immigration system should not treat EU workers favourably at the expense of non-EU workers. Rather, such a system would remove the free movement principle – seen as creating an uneven level playing field, and replace it with a system, which rewards high skills and earning potential while restricting low-skilled/low-paid workers. The mobility of lower skilled/paid workers is not to be managed by a particular policy route (with a possible exception for seasonal workers) because there is an 'existing stock' of such people

(MAC, 2018: 5) which could and should be utilised. What is noteworthy here is that despite its attempt to be value-neutral and technocratic, the MAC report articulates a rather ideological misunderstanding of the EU's freedom of movement principle. The report refers to it as a 'policy choice' (MAC, 2018: 110) which allows (mostly) lower-skilled migrants from newer member states to migrate to the UK without any restrictions. This is seen to be a violation of the EU principle per se, as well as the UK's ability to apply differentiated policy filters to migration flows. Equating free movement with unrestricted migration is not just factually incorrect, as this chapter has already established, but it also overrides empirical evidence and legitimises further restrictive migration measures that divide migrants into being more or less valuable.

From an empirical perspective, Brexit was an important event because it triggered a rather unprecedented obsession of counting EU nationals in the UK. The sudden realisation that mobile EU citizens hadn't been – before Brexit, that is – migrants in the legal sense of the term due to the unique characteristics of EU supranational citizenship (see here Arcarazo and Wiesbrock, 2015), opened up a metaphorical Pandora's box as it gradually emerged that thousands of EU nationals from the core member states who had lived in the UK for decades had never bothered to register their UK residency or to acquire British citizenship because they didn't really have to. Therefore, in order for any post-Brexit migration policy to be viable, it had to first take stock of how many EU citizens lived in the UK. Migration data indicates that as of 2019, an estimated 3.7 million EU citizens lived in the UK, 47 per cent of which came from the so-called EU-14 countries, or the core member states (Vargas-Silva and Walsh, 2020). According to the UK government's EU Settlement Scheme Statistics, 5.26 million applications for settled and pre-settled status were logged by EEA nationals (Home Office, 2022).

In this respect, post-Brexit migration policies sought to reconcile the technocratic and mythical, securitised understandings of the value of migrants by introducing an Australia-inspired points-based immigration system. Here, it is worth reiterating that despite claims of novelty, a points-based immigration system was previously introduced by New Labour. As much as pre-2016 political manifestos and policies have focused exclusively on non-EU migration, post-Brexit migration policies disproportionately affect EU nationals already living in the UK, as well as Europeans who are yet to move here. The new immigration system rejects any continuity or adaptation of the freedom of movement principle and aims to return the prerogative of granting access to the country to UK authorities.

On the one hand, the new system differentiates in legal and policy terms between those who arrived in the UK before the end of the transition period on 31 December 2020, and those who might arrive afterwards. The former are required to obtain either a pre-settled or settled status, which guarantees the retention of pre-Brexit status and rights. The latter become subject to the new points-based system and will have to apply for visas and work permits if they meet the set thresholds in terms of skills, language proficiency and salary.

On the other hand, it equalises the treatment of EU and non-EU nationals who wish to come to the UK. While the new system alleges that it won't discriminate on the basis of nationality, it has a de facto bias against people who are not of working age, economically active, or with significant earning potential. Empirical research demonstrates that the lines of work where people typically do not earn above the official annual threshold of £25,600 are hospitality, retail, care and food processing (Fernandez-Reino et al, 2020). These are all sectors with a significant number of migrant workers, many of which from Eastern EU member states.

Lessons (un)learned

This chapter has put forward the argument that despite different styles of executive power and rhetoric, migrant workers have been historically and politically treated and perceived as robots in the sense that their main defining characteristic, and sole point of redemption, is their physical and mental ability to be economically productive. At the same time, their humanity – the possibility that they get ill, unemployed, pregnant, and so on – is not only denied, but deemed politically and socially problematic. Throughout the examined period (from the early 20th century to the post-Brexit period), different governments sought to provide different ideological justifications for restrictive and/or selective immigration policies. From a more explicit focus on racialised preferences and concerns about cultural proximity and differences, to a more technocratic and managerial approach to migration, which sought to align labour mobility with the needs of the British economy, to a distinctly securitised approach dismissive (or at least sceptical) of evidence and preoccupied with anxieties about the alleged threats that migrant workers pose not only to the British economy, but to British society as well.

This continuous dehumanisation is carried out through securitised media and political discourses, and austerity-motivated policy restrictions against the backdrop of an evidence-adverse context. Thus, despite their differences, the racialised, the technocratic and the securitised approaches to economic migration are unified by their common crude utilitarian and dehumanising understanding of migrant labour.

6

Expensive Robots versus Cheap Migrants

The inevitability of migration and the necessity of cheap labour

As the previous chapter noted, migration policies are embedded within different ideologies and their adoption and implementation depend on competing political agendas. It was also noted that regardless of whether a specific migration policy could be understood as racialised, technocratic or securitised, there is an underlying assumption that there is (some) labour value to be extracted from the body of the migrant. Here, it is important to note that the dehumanisation of migrant workers is not the result of the type of labour that they do, rather, it is the result of political and media rhetoric and policy praxis, which effectively disenfranchises migrant workers. At the core of all these approaches to migration policy lies the understanding that *some measure* of migration is inevitable and essential for the British economy. The general availability of foreign-born people who are willing and able to work – first through the Commonwealth and later on via the UK's membership of the EU, has meant that for successive Conservative and Labour governments, migration policy has been an auxiliary tool for fiscal and social policies. In other words, there is a tacit consensus across the political spectrum that there is economic utility in migration, which cannot be adequately substituted either by the existing native labour supply, or by technological advancements, including but not limited to robots and artificial intelligence (AI).

The current chapter argues that employers in sectors with a high concentration of migrant workers are most likely to continue to rely on such precarious migrant labour, despite pre- and post-Brexit promises for increased investment in automation in labour-intense and migrant-dominated sectors of the economy, such as agriculture. The main reasons for this continuous dependency are as follows. First, despite the general fascination with the idea of automation, especially within the context of Brexit, no government to date has committed to properly invest in it due to its high costs – politically, socially, as well as in monetary terms. Such a commitment will also require a more substantial and nuanced understanding of the role of the state in regulating intra-societal interactions. Whether the intensification of automation developments creates a broadly positive, or a broadly negative societal impact, such intensification will inevitably transform racialised hierarchies within society in general and the labour market in particular, exacerbating inequalities in doing so. Thus, some mitigation will be essential in the form of more generous social policies, that is, subsidised further education, reskilling and retraining, and/or Universal Basic Income. In the aftermath of the 2008 financial crash and the ongoing and entrenched discourse and praxis of austerity, no government has expressed a firm commitment to manage and mitigate such societal transformation should there be a more systematic automation rollout. Further, the analysis in the chapter will demonstrate that the financial costs of developing as well as accelerating automation developments are to be borne exclusively by employers and that such costs could be prohibitively high for smaller enterprises. The ongoing political and policy uncertainties around Brexit and COVID-19 are also likely to prolong the atmosphere of ambiguity and non-commitment.

Second, and related, the only viable remedy to the chronically low productivity in the UK will continue to be provided by migrant workers. By examining specialist reports, political and media statements, as well as the Pick for Britain campaign, the chapter will unpack and critically engage with the trope of *cheap labour* with a specific reference to mobile EU workers. The neoliberal and largely hands-off approach to intra-EU labour migration on the EU level, as well as on the local, national level in the UK (before Brexit, that is) have contributed to an understanding of low-skilled/low-waged migrants as a low-maintenance, disposable, easy to hire and easy to fire workforce. Such understanding has been facilitated by growing intra-societal inequalities and a general stagnation of wages in some Eastern EU countries, which has contributed towards an increased volume of East–West intra-EU mobility and, consequently, an increased availability of workers. In other words, the essential financial and training commitments towards algorithms, robots and AI is by no means exercised towards their human counterparts, which effectively makes the latter the preferred, cheaper alternative to automation.

The chapter begins by unpacking the entrenched and somewhat infamous trope of cheap labour and its political and policy manifestations. Specifically, we explore how the freedom of movement principle has converged the desire to migrate with the actual ability to do so, as well as the significance of the Posted Workers Directive (PWD) for intra-EU labour mobility. We then move on to discuss the role of anti-EU and anti-EU migrants rhetoric for Brexit and the plethora of political promises around the 2016 referendum and afterwards to reduce the UK's alleged over-reliance on migrant labour and to reinvent the image of the country as a progressive global high-tech powerhouse. Such claims and promises are evaluated against the backdrop of key automation reports published between the Brexit vote in 2016 and the end of the transitional period in December 2020. Finally, the chapter will explore the links between COVID-19 and migration, particularly in terms of the introduction of (some) automation by stealth during the global pandemic and its impact on employment and existing inequalities, as well as the effective degradation of essential workers to a social class term where migrants are praised and vilified in equal measure.

What do we mean by *cheap labour*?

Cheap labour has often been associated with foreign labour and, since the mid-2000s, it has been used extensively with regards to Eastern EU workers coming to the UK under the EU free movement principle. By the end of the Brexit transition period in December 2020, 55 per cent of migrant workers were EU-born (Sumption, 2021: 5). Due to important differences between the permissive nature of the free movement of persons principle and the more restrictive and highly regulated nature of migration policy governing non-EU mobility, EU-born workers were more likely to work in low-waged, low-skilled and short-term/precarious work compared to non-EU migrants, as well as British-born ethnic minorities (McDowell, 2009; Pencheva, 2021). Following the three rounds of Eastern EU enlargement, mobile Eastern EU citizens tended to be concentrated in the low end of the labour market, particularly in sectors such as agriculture, hospitality, construction, food and drink processing (Rolfe et al, 2013). Indeed, the availability and flexibility of Eastern EU workers has been highly beneficial for employers (Rolfe et al, 2013; Rolfe, 2017).

The trope of cheap labour is particularly susceptible to politicisation because it is difficult to define and measure. Fernández-Reino and Rienzo (2022) have demonstrated empirically that by the end of the Brexit transitional period 48 per cent of workers born in North America, Australia and New Zealand, and 47 per cent of workers born in India were in high-skilled jobs, as well as 45 per cent of those born in EU-14 countries. However, workers born in Eastern EU member states were overrepresented in the low-skilled end of the labour market, doing jobs such as cleaning, packing or waitressing. This applied to 22 per cent of EU-8 nationals and 19 per cent of EU-2 nationals (Fernández-Reino and Rienzo, 2022). The same report also highlights the fact that employees born in post-2004 EU countries also had the lowest earnings dispersion of all the foreign born, whereas workers from the US and Oceania were the highest earners (Fernández-Reino and Rienzo, 2022). Markova and Black (2008) have noted that post-2004 labour migration is also characterised by a substantial mismatch between formal qualifications and type of employment. Defining and measuring cheap labour is further complicated by the fact that it isn't just formal qualifications that define a certain job as low-paid and/or low-skilled. Soft skills, such as cooperative behaviour, communication and people skills, are essential for low-paid jobs in social care and patient-facing roles in the National Health Service (NHS; see here Anderson and Ruhs, 2012). Physical strength and agility are a must for agricultural work, as well as for some types of physically demanding jobs in retail, construction, food and drink processing, although these qualities are not deemed to reflect skills per se.

The common usage of the phrase *cheap labour* is predicated on the assumption that migrant labour is cheaper, that is, paid less than the national resident workforce and that migrant workers contribute towards a race to the bottom by being willing to accept lower wages. Such arguments are advanced by scholars (Kvist, 2004; Caro et al, 2015) who emphasise that the difference in living standards between newer and poorer EU member states and core and wealthier member states is effectively undermining the existing industrial compromises in the latter (see also Andrijasevic and Sacchetto, 2016). Kvist (2004) also notes that the migratory surplus in Eastern EU member states is more substantial compared to that of previous enlargement rounds, and that it had been artificially suppressed by restrictive mobility regimes during and after the Cold War era. This specificity, according to Kvist, has contributed to the larger than anticipated influx of Eastern EU nationals following the 2004 and 2007 enlargement rounds. It is important to note, however, that the social dumping argument is also entrenched in political and media rhetoric where it takes the form of crude descriptive statements about the national minimum wage in an Eastern EU country, which is then crudely juxtaposed to the British equivalent, often with little if any regard for context. Some examples of such discursive practices are presented in the empirical section that follows.

A key factor behind the political and public prominence of the *cheap labour* trope was the PWD and the various ways in which it was (ab)used. The PWD 96/71/EC is an EU directive concerned with the free movement of workers within the EU. It defines a posted worker as an employee who is sent out by their employer to carry out a service in another EU member state on a temporary basis (full text available via ec.europa.eu). The literature is exclusively focused on post-2004 East–West dynamics because prior to the Eastern EU enlargement labour mobility between member states in general, and via the provisions of the PWD in particular, has been limited and socially unproblematic. Thus, the scholarship identifies several key issues associated with the inappropriate and/or *alegal* use of the directive: wage dumping, unfair labour market competition, promoting practices of outsourcing and subcontracting, as well as the general difficulty of unionising posted workers. Alberti and Danaj (2017) have argued that in Britain neoliberal deregulatory policies and practices of the 1980s and 1990s have been conducive to the creation of an environment where employment agencies and subcontractors de facto control the labour market. Their findings suggest that transnational labour mobility in conjunction with a largely independent and deregulated private sector perpetuate the need for cheap temporary workers as it strips them of their rights and excludes them from institutional provisions for equal treatment, that is, welfare provisions.

Along similar lines, Barnard (2009a, 2009b) argues that the (ab)use of the PWD benefits disproportionately businesses in accession states at the expense of labour laws and collective bargaining provisions in the wealthier, core member states. Cremers et al (2007: 524) in their social and political reading of the directive and its impact, postulate that 'from the very beginning, the question whether protection of the posted workers stood on an equal footing with, or was subordinated to, the free provision of services was at the centre of the struggle to establish the directive'. The article points out that the three rounds of Eastern EU enlargement (2004, 2007 and 2013) have posed serious challenges for many core member states as the migration flows from East to West increased. The authors also note that whether or not the implementation of the PWD leads to social dumping depends largely on how labour relations and collective bargaining are set up and regulated in some of the wealthier member states. For example, in countries with high labour costs and decent regulation/standards (Scandinavia, Netherlands, to some extent Germany and Austria) the social dumping effect is likely to be minimal or non-existent, whereas in countries with more deregulated labour markets, weaker unions and overall low labour costs (the UK, Ireland, Portugal), social dumping will be more pronounced as well as politically and socially problematic (Cremers et al, 2007; see also Alsos and Eldring, 2008; Berntsen and Lillie, 2016 among others).

The other side of the cheap labour debate, or the so-called integrationist position, postulates that the free mobility of Eastern EU workers is beneficial for everybody, that is, for countries of origin and destination, as well as for the mobile individuals themselves (Bonin et al, 2008; Kahanec et al, 2010). Research has indicated that increased East–West intra-EU mobility had overall neutral or positive effects on the EU economy (Kahanec and Zimmermann, 2010), on GDP per capita, employment rates and inflationary pressures (Kahanec et al, 2013), as well as on welfare systems (Dustmann et al, 2003). In other words, it is a win–win scenario where the Eastern European countries of origin effectively rid themselves of a labour surplus that cannot adapt to their changing economies, whereas destination countries in Western Europe manage to fill in vacancies for jobs across the sectors of agriculture, manufacturing and services without necessarily having to invest in reskilling or further education for the native workforce (Anderson and Ruhs, 2012). Politically, the notion that the pool of available workers from accession states is both vast and cheap has informed the rhetoric and praxis of the New Labour era and has effectively reduced the liberal defence of immigration to the economic utility of migrant workers. In other words, if migrants are willing and able to do the jobs that *we* don't want to do for wages that *we* won't accept, then (some) migration is acceptable.

While we do recognise and appreciate this debate between those who claim that Eastern Europeans contribute towards wage dumping and those who praise the flexibility (and precarity by association) of post-accession labour markets, we do not necessarily think that such a dichotomous understanding is helpful for the empirical case at hand, namely intra-EU mobility. First, while the political realities of intra-EU mobility reaffirm state-centrism as the main prism through which threats and vulnerabilities are discursively constructed, this mobility is characterised by multidirectionality and a sense of temporal flexibility. It is often portrayed as uncontrollable and politically problematic because it operates within a set of supranational permissive and enabling policies with regards to mobility within and between EU member states, which is perceived as disempowering the British state in its essential role as a guardian and protector of the nation. To wit, the freedom of movement principle could be conceptualised as a form of permissive mobility regime, which, albeit with some conditions and restrictions, aims to facilitate rather than to hinder mobility.

Second, and related, the EU accession of Eastern European countries has resulted in the convergence between people's desire to migrate and their ability to do so. This is referred to as the aspiration-ability model (Carling, 2002; Carling and Schewel, 2018) and is based on the idea that human mobility should be analysed against the backdrop of restrictive migration policies and regimes because having the desire to work abroad does not necessarily mean that one would be able (or, indeed, permitted) to do so. For many Eastern Europeans gaining EU citizenship rights made cross-border mobility an affordable, risk-free and visa-free enterprise. Despite widespread differentiated treatment of Eastern EU nationals in the form of up to seven years of transitional arrangements, millions have been able to try their luck abroad.

Third, this multidirectional and largely unobstructed mobility creates conditions for precarious labour, which is *cheap* insofar as the process of bordering has shifted from policing physical borders to more subtle processes of institutional inclusion and exclusion. So, while by virtue of their EU citizenship (until Brexit that is), Eastern Europeans could navigate the British labour market more freely than without it, Ciupijus (2011) argues that Britain has been reluctant to extend full social rights to these workers. As a result, many have been confined to low-waged, temporary and/or precarious work. At the end of New Labour's technocratic approach to migration and the advent of a securitised approach and a hostile environment, EU workers' rights to social benefits and legal aid have been gradually restricted. Their inclusion in the UK labour market has been successful on account of their high levels of employability and alleged strong work ethic, but also limited and conditional on the institutional level. The libertarian defence of migration on the basis that migrants' fiscal contributions outweigh any potential claims they might put on the social purse often conveniently omits that most claims on the social purse are either not possible, or severely restricted.

Thus, we seek to advance the understanding that yes, EU workers are *cheap labour* because they have been discursively constituted as such in political, policy and media narratives. Both migration and labour automation policies de facto make migrants preferable (read *cheaper*) to national residents for the following reasons. First, prior to Brexit EU citizens operated within a permissive immigration regime afforded by the free movement principle. This laissez-faire approach was favoured by the EU and was implemented (albeit with some hesitation) on the national UK level. While this formal equality of opportunity has enabled mobile EU citizens from accession states to navigate the British market much freer and easier than without it, this has contributed towards the political, economic and cultural entrenchment of precarious labour which renders their economic inclusion limited and conditional. Second, and related, the lengthy transition period post-2016 and the ongoing uncertainties and ambiguities regarding post-Brexit immigration policies are likely to further exacerbate the issue with cheap labour. The Australian-style points-based immigration system poses restrictions and demands on post-Brexit labour mobility for both workers and employers and while the new annual earnings threshold of £25,600 might be achievable for IT or financial services professionals, essential workers in retail, agriculture, social care and the NHS are unlikely to meet it or to be offered a visa sponsorship, which in turn will open a Pandora's box of workers' exploitation in the years to come. The seasonal agricultural scheme is currently in its third year of ad hoc non-committal extensions, despite unobjectionable evidence and pleas from farmers and growers that labour shortages in the agricultural sector are severe.

Finally, the discourse around automation in the aftermath of the Brexit referendum, and within the broader context of populist and anti-immigration rhetoric, has been marked by a fascination with the idea of automation while remaining remarkably ambiguous in terms of the necessary practical commitments. In addition, the insistence that automation will be prohibitively expensive and will have a measurable impact on pricing, purchasing power and social inequalities that transcend electoral cycles, results in the kneejerk reaction of successive governments to continue kicking the issue down the road while reinforcing the idea that there is an available *cheap* alternative: the *men machines*.

Brexit: a technologically advanced post-immigration utopia?

The pre- and post-Brexit political climate was marked by promises and claims that 'taking back control' will reduce the alleged over-reliance of the British economy on migrant workers and will boost productivity and growth via the adoption of new technologies. Then food and rural affairs secretary Andrea Leadsom told farmers at the 2017 National Farmers' Union's annual conference in Birmingham that instead of complaining about ongoing structural difficulties to recruit – and retain – tens of thousands of seasonal workers from the EU, they should instead invest in technology and machinery in order to boost their productivity in the long term. She added that 'we must not forget that a key motivating factor behind the vote to leave the EU was to control immigration' (quoted in Butler, 2017). Speaking at a 2017 Conservative party fringe meeting, Leadsom – an adamant Leave supporter – claimed that she was 'so excited' on behalf of those who would 'live most of their adult lives outside of the European Union' (in Forrester, 2017), adding:

> [F]or me as one of the key champions of leaving the EU this is about you. Sometimes people say older people voted for leave, everybody young wanted to stay, but that is in my case absolutely not the case – I was a proponent of leave for your sake, for my kids' sake, for the next generation. For some of you it may feel scary but for me, on your behalf, it's really exciting. (In Forrester, 2017)

There are two main issues that need to be unpacked here: the first one relates to the alleged over-reliance on migrant workers and the importance of 'taking back control', the second concerns the post-Brexit efforts of the political establishment to create facilitating conditions for the gradual replacement of migrant labour by machines.

Empirical research (Blinder and Richards, 2020) has demonstrated that between 2001 and 2016 immigration was the number one concern for 56 per cent of the British public. In 1994, the starting point of this data series, less than 5 per cent of respondents thought of immigration as a concern, and it remained rarely mentioned prior to 2000. In the year or so before the EU referendum, between June 2015 and June 2016, immigration was consistently named as the most salient issue facing the country, peaking at 56 per cent in September 2015 (Blinder and Richards, 2020). Based on polling and survey data Blinder and Allen (2016) also demonstrate that public concerns about immigration became more prominent around the mid-2000s and that a significant factor for these concerns is the perception that EU migration is uncontrollable and impossible to forecast and plan for. The evidence indicates that the British public is concerned about EU as well as non-EU migration and wishes that fewer people were let in and under stricter conditions. Related to this point, Blinder and Allen show that immediate before the 2016 referendum, the public opinion was that the economic and social costs of intra-EU mobility significantly outweigh its (fiscal) benefits (Blinder and Allen, 2016: 7; see also Dustmann and Frattini, 2014).

A few words should also be said here about the role the British media played in terms of its overall approach to migration coverage and its significance for Brexit. British print media has a rich history in reporting on migration-related issues and has followed closely all rounds of Eastern EU enlargement (Mawby and Gisby, 2009). However, Allen (2016) convincingly argues that journalists tend to actively compete with political leaders and other actors in framing migration debates instead of being content with reporting on others' views on migration policies. Furthermore, the increased politicisation of migration coverage, the tendency to play fast and loose with the term 'migrant' (see here Anderson and Blinder, 2015), and the growing editorial practices of either using evidence rather selectively, or dismissing it altogether, has resulted in a national migration coverage that is hostile and toxic. Michael Gove's infamous line that people have had enough of experts speaks to a securitised discourse which not only dismisses empirical evidence and undermines professional expertise, but it also de facto validates public perceptions and anxieties. To wit, it has been noted that the quality of debate in British news media over Eastern European nationals coming to the UK has been questionable and that it has fuelled anxieties over the possibility for yet another mass influx of Eastern European nationals (Mawby and Gisby, 2009; Vicol and Allen, 2014; see also Balch and Balabanova, 2016). Balch and Balabanova (2016) point out that British news media's arguments about migration from Bulgaria and Romania could be interpreted through the prism of 'moral panics', while Pencheva (2020b) has argued that print media narratives on Eastern EU workers have been discursively securitised. Eastern EU migrants have been portrayed as cheap labourers and accused of stealing jobs, undercutting wages and of putting pressure on the NHS, schools and other welfare services (Pencheva, 2020b).

The gradual but steady conflation between 'Europe' and 'EU migrants' against the backdrop of a traditionally Eurosceptic British media environment (Daddow, 2012) made intra-EU mobility a significant factor for Brexit. In terms of political communication, both the Leave and Remain campaigns reflected substantial misunderstandings over what the EU is and the nature of the relationship between the UK and the EU. Both campaigns effectively fetishised British sovereignty (Pencheva and Maronitis, 2018) as an essential and non-negotiable aspect of British identity while the Leave campaign was also praising its emancipatory potential should the public decide in favour of leaving the EU. No more daunting and demanding EU regulations, no more supranational criticism on issues of social policy and human rights. Key Conservative politicians at the time were claiming that cheap labourers from Eastern Europe should and must be replaced by an increased investment in automation technologies. Andrea Leadsom said that there is 'a whole raft of new technologies that complement the workforce' (Butler, 2017). Then Chancellor of the Exchequer Philip Hammond has said that 'there are steps of automation that can be taken by investing capital but are not taken when access to low-cost labour is available', while Conservative MP Alan Mak claimed that AI and mass automation should be cornerstones of British industry after Brexit (Condliffe, 2017).

Whether or not the UK economy has a structural dependency on migrant labour is an important question, but it does not necessarily have a straightforward answer. On the one hand, there are arguments that overly lax migration policies contribute towards the creation of a bottomless pool of skilled and willing workforce, which effectively desensitises UK employers from improving working conditions and pay (Anderson and Ruhs, 2012). Some of these arguments were explicated in greater detail in the previous chapter but what is crucial here is that the existence and size of labour shortages are not a given – as much as both (New) Labour and Conservative policies seem to suggest; rather, these are contingent upon employment conditions, pay and the general availability of resources to upskill and reskill the British workforce (Anderson and Ruhs, 2012). In fact, Anderson and Ruhs argue that more often than not, jobs that are officially classified as low-skilled actually require a range of soft skills and competences that disadvantage the British workforce at the expense of foreign-born workers. Examples here include approachability and friendly demeanour in customer-facing jobs, compliancy, self-discipline and cooperation (Anderson and Ruhs, 2012: 25). This could easily result in a structural demand for a workforce which could be extensively controlled by employers.

On the other hand, there are the arguments that migrants often possess a superior work ethic compared to the local workforce and that they are willing to work anti-social hours doing jobs that need doing but that nobody else wants to do (Blanchflower et al, 2007; Markova and Black, 2008). Indeed, the notion that migrant workers have a superior work ethic and high levels of employability gained further traction following the rounds of the Eastern EU enlargement, which created new opportunities for those actively seeking economic security for themselves and their families. Free movers practise an understanding of equality that is now the benchmark by which social inequalities are perceived: formal equality of opportunity (Faist, 2013), the pursuit of which is made possible by a set of enabling circumstances, that is, liberalisation of immigration policies, visa and labour market regulations, which allow EU citizens who live and work in other member states to contribute to a Europeanisation from below (Recchi and Triandafyllidou, 2010). The aim of this mobility is to assume some of the functions that traditionally are seen as primary to the state: providing basic economic security for its citizens. Academic research has also pointed out that workers from Eastern EU member states demonstrate higher than average levels of mobility compared to other EU nationals (Black et al, 2010; Sanchez, 2015). According to the 2010 EU Labour Force Survey, Bulgarian and Romanian nationals of working age (15–64) have significantly higher employment rates compared to the working age populations of countries, such as Italy and Great Britain (European Commission, 2011).

In a study of mainstream British media's portrayal of Bulgarian and Romanian workers, Pencheva (2019) argues that the trope of stealing jobs has been elevated to an unique blend of securitisation and racialisation whereby Bulgarian and Romanian workers' work ethic and high levels of employability are simultaneously praised by the British press vis-à-vis those of the White British majority, as well as British ethnic minority workers; while at the same time accusing them of unlawfully and unscrupulously scamming the local workforce out of employment opportunities and decent wages. The study suggests that such a paradoxical portrayal was characteristic for both tabloid and broadsheet editions. Further, while media and political discourses recognise the problem with precarious and exploitative employment, more often than not Eastern EU workers were explicitly blamed for their own precarity – if they weren't stealing jobs and undercutting wages, they wouldn't be exploited in the first place (Pencheva, 2019, 2020b; see also Andrijasevic and Sacchetto, 2016).

The study gives an example from an article from the international pages of the *Sunday Telegraph* which praises the work ethic of Bulgarians and Romanians against the backdrop of the anxiously anticipated influx from the two countries. The text claims that 'British firms like Bulgarian staff. They are quiet, loyal workers who don't complain or cause trouble'. This sentiment appears to be shared by a Romanian builder in Bucharest who is quoted saying that 'on British building sites you only see foreign nationals, go into a pub, though, and all you'll see is the English' (Freeman, 2006: 27).

Further, the notion that the native British labour force is work-shy and overly reliant on welfare payments is reiterated by another news article from the *Telegraph*. The article explains the high employability of Bulgarian and Romanian nationals by referring to their allegedly better work ethic. The newspaper claims that 'Bulgarians are preferred because they are hard workers' and goes on to suggest that 'they are not going to the UK for pleasure, they are going because they need money and they are willing to work for it' (Evans, 2013: 1).

Daily Mail columnists James Slack and Becky Barrow also focus on the viewpoint of business owners. The article quotes the chief executive of Domino's Pizza, Lance Batchelor, who claimed that it was 'harder to hire staff in the UK' and 'criticised the government for making it more difficult to recruit immigrants instead' (Slack and Barrow, 2013: np). The tabloid also quotes the chairman of the online delivery firm Ocado, Sir Stuart Rose, who praised the work ethic of Eastern Europeans by saying it is their own choice to work long hours for little pay and that such a decision should not be criticised because it is good for business. He claimed that:

> It is wrong to criticise immigrants who are prepared to work for lower salaries that Britons turn down. These people are entitled to come here and if these people want to come here and work the hours they are prepared to work for the wages they are prepared to work, so be it. There is nothing against that. (Slack and Barrow, 2013: np)

That said, the alleged high employability of Eastern EU migrants and their superior work ethic have also been racialised in media, policy and political discourses (Fox et al, 2012; Pencheva, 2019). On the one hand, the praise of migrant workers for getting jobs and for putting up with low pay and often poor working conditions has positioned them as superior vis-à-vis the alleged work-shy Brits (Skeggs, 2004). On the other hand, these seemingly positive qualities have been seen as amplifying structural problems with lack of employment opportunities and social mobility for British ethnic minorities (Pencheva, 2021). It is worth noting that although Central and Eastern European citizens exhibit higher levels of labour market mobility and employability compared to members of established ethnic communities, their work patterns are marked by job insecurity, low pay and precarious recruitment practices. In comparison, McDowell (2009: 750) argues, British ethnic minorities often struggle to get employment opportunities or tend to be in long-term but low-pay employment.

As far back as 2006, *Times* columnist Carl Mortished argues that, despite public criticism over the large influx of Poles, John Reid's decision to introduce work restrictions for Bulgarians and Romanians has an 'ulterior motive' (Mortished, 2006: 60). The columnist suggests that Bulgarians and Romanians are socially unproblematic, explicitly referring to their putative Whiteness and good work ethic: 'these immigrants from the East are easily absorbed, learn English quickly and pass unnoticed. They don't seem to want to kill us and they work like Trojans for whatever money they can get'. Mortished further argues that Mr Reid is so 'fearful' of the 'new Europeans' because their successful labour market integration will expose the unsuccessful societal integration of 'young Asian males', who are particularly susceptible to radicalisation. Bulgarians and Romanians, he claims:

[R]epresent competition for Britons seeking unskilled or semi-skilled jobs. And it is black and brown Britons that are particularly threatened by the pale-skinned workers from the Baltic and Black Sea. Unemployment among ethnic minorities in Britain is disproportionately high. There is massive concern within government about joblessness among young Asian males and genuine fear of what idle thoughts will fill those years of empty days. (Mortished, 2006: 60)

As the foregoing sub-section has indicated, there is a strong argument that many employers, especially those in the low end of the labour market, will express a preference for migrant workers. Eastern EU workers are a case in point because the freedom of movement principle enabled labour mobility by making it affordable (that is, removing visa and/or work permits costs) for both workers and employers (Ciupijus, 2011). Despite some undeniable benefits afforded by the EU freedom of movement principle, the quality of media discussions, as well as political and policy choices such as restricting EU workers' access to welfare and public services, has disproportionately benefited employers (whether they are decent or unscrupulous) at the expense of migrant workers who (un)knowingly have come to bear all the costs and risks associated with their mobility.

At the time of writing, it has been just over five years after the so-called Brexit referendum. Net migration continues to be in the hundreds of thousands – despite the set of hostile environment policies that aimed at bringing it down to the tens of thousands, many EU citizens are leaving the UK, with fewer and fewer choosing the come here after Brexit. The Office for National Statistics (ONS) estimated that in the year to June 2017, UK net migration stood at 230,000, which is 106,000 lower than it was the year before (ONS, 2017). While still relatively high overall, three-quarters of this substantial dip in net migration levels has been attributed to the exodus of UK-based EU nationals around 2016 and immediately afterwards (Condliffe, 2017). There are ongoing uncertainties around post-Brexit migration policies – especially for those workers at the low end of the labour market who are unlikely to meet the annual salary thresholds under the post-Brexit points-based immigration system, as well as the looming possibility that the hostile environment might affect the residency rights of EU nationals. Furthermore, there continues to be a gap between the emancipatory political rhetoric on the necessity of replacing cheap migrant workers with automation and technological advancements, and the empirical reality of lacking commitment and investment in the latter.

Examining a few official reports on automation produced between 2016 and 2020, that is, between the Brexit referendum and the end of the transitional period, reveal that despite political promises about the reinvention of the UK as a technologically advanced global nation, the current state of automation is perhaps best described as *much ado about nothing*. Apart from the anti-EU workers impact of Brexit, low-skilled and low-paid work continues to dominate in the UK economy, which is more vulnerable to specific policy choices, particularly austerity, than the general threat of automation (Spencer and Slater, 2020: 118). Austerity, or the decision to reduce public spending, is said to be a more important factor for a long-term stagnation of investment, growth, productivity and wages than automation and AI (Spencer and Slater, 2020).

A 2017 Royal Society for the Encouragement of Arts, Manufactures and Commerce (RSA) report (Dellot and Wallace-Stephens, 2017) lends further support to the argument that the contemporary state of automation in the UK could be best described as patchy. The empirically rich report suggests that in the UK, the anxiety about automation is significantly higher than any actual investment in it. The authors make the important point that AI and automation technologies are complex and multidimensional phenomena that are not predetermined to produce a particular outcome because they remain embedded within social and political contexts, and as such are subjected to ideologically motivated policy choices. While admitting that case studies and empirical data lend support to the statement that risks are skewed towards low-skilled/low-waged professions, the report reveals that the intersectoral vulnerabilities within this category are not equally distributed. The potential to automate job-related tasks in logistics and retail are substantially higher than in hospitality and leisure (4 per cent), medical and health services (2 per cent) and in education and care (3 per cent) (Dellot and Wallace-Stephens, 2017: 6).

A recent report on migrants' participation in the UK labour market confirms that as of 2019, migrants were overrepresented in the hospitality sector (30 per cent), transport and storage (28 per cent), communication and information technology (24 per cent) and health and social work (20 per cent) (Fernández-Reino and Rienzo, 2021). This means that any artificially induced labour shortages, that is, by clamping down on EU labour migrants due to perceived populist pressures, cannot be easily automated due to the manual and interpersonal skills that are needed to clean a hotel room or to provide social care to old and/or vulnerable patients (Condliffe, 2017). Further, such a replacement could be prohibitively expensive. Any impact of automation and AI on jobs, as well as consequent policy choices related to taxation, welfare, regulations and/or further education will only be relevant if and when there is significant investment and automation technologies and if and when these are rolled out and implemented. According to Dellot and Wallace-Stephens (2017: 7):

> Our RSA/YouGov poll finds that just 14 percent of business leaders
> are currently investing in AI and/ or robotics, or plan to in the near

future (the figure is just 4 percent for small businesses). Many think that the technology is too costly or not yet proven. For others, concepts such as machine learning, deep learning and cloud robotics appear to be completely new.

The 2019 cross-party Business, Energy and Industrial Strategy Committee (BEIS) report provides a more empirically grounded and rhetorically measured take on the issue of automation of work. The report makes it clear that there isn't a clear government strategy for supporting businesses in adopting labour automation technologies, especially after the abolition of the Manufacturing Advisory Service. Businesses are often unclear on how and where to seek support and essential dialogue with academia and trade unions, which could facilitate meaningful and long-term engagement of workers via exploring options for lifelong learning and reskilling. Without over- or underplaying the risk of job losses, especially in the low end of the labour market, which would disproportionately affect younger workers, the report is clear that the adoption of industrial automation is imperative if the UK is to remain a competitive global economy. Having a sustainable, long-term strategy and cross-sectoral collaboration is essential not just for boosting productivity and for improving the quality of work life, but also as a way of protecting jobs from being outsourced (BEIS, 2019).

The official government response to the BEIS report was published on 20 March 2020 on the website of the UK Parliament. Overall, the response could be described as extensive, agreeable yet remarkably vague on the practical level. It is evident that the official government position is that automation is – generally speaking – a good thing and that there is fertile ground for its adoption in the UK with its world-class universities and research facilities, and its dynamic service sector. To wit, the prepared statement claims that the government has strengthened its focus on Robotics and Autonomous Systems (RAS) with the launch of the Robotics Growth Partnership and a national ambition to 'put the UK at the cutting edge of the smart robotics revolution ambition, turbo-charging economic productivity and unlocking benefits across society' (House of Commons, 2019). The government response outlines various commissioned reports, pilot schemes and initiatives, as well as tax relief for Small and Medium Enterprises (SMEs) that should facilitate the feasibility of adopting digital technologies and automation.

In terms of identifying new areas of automation for further waves of Industrial Strategy Challenge funding, the government highlights the agricultural sector as an area of notable activity. Specifically, technological advancement in this sector is seen through a green/sustainable lens, as well as a way of boosting productivity. The text suggests that 'a range of robotic solutions are emerging in the food and farming to improve efficiency, manage seasonal labour shortages and reduce carbon', as well as stating that 'investment in automation technologies across the Agri-Food chain to help improve productivity, provide safer work environments, reduce waste and improve product quality' (House of Commons, 2019: np). More recently, the first 31 projects of the £90 million Transforming Food Production were announced in June 2019, combining AI, robotics and earth observation to improve supply chain resilience in the agri-food sector and plans for post-Common Agricultural Policy Future Farming arrangements were announced, including research and development funding schemes to facilitate collaboration and pioneer innovative and efficient farming techniques (House of Commons, 2019).

While the official government discourse is marked by optimism and confidence, these do not appear to be shared by farmers and growers. Meurig Raymond, president of the National Farmers' Union, told the *Guardian* that while it was theoretically possible that robots could help farmers reduce their reliance on (foreign) labour, horticulture remains a 'people-based business' and driverless tractors or mechanical harvesting simply cannot go far enough to substitute human labour (Butler, 2017). Furthermore, any meaningful transition towards a more automated mode of production will require significant up-front costs and investments for farmers, as well as government support via low-interest loans and tax allowances. Raymond assessed the latter as 'simply insufficient' (Butler, 2017). There is also the need to guarantee prices and to have a sense of predictability and security, which has been particularly challenging in the aftermath of Brexit.

However, a utopian futuristic vision of a post-Brexit Britain where working smarter rather than harder would be possible due to automation technologies and AI – all while the unwanted xeno Homo Oeconomicus is kept at bay – has not materialised yet. Post-Brexit reports demonstrate a distinct lack of government commitment to investing in labour-optimising technologies and to supporting farmers, as well as small businesses in adopting such technologies (Dellot and Wallace-Stephens, 2017; Gilbert and Thomas, 2021). In fact, in light of the devaluation of the pound sterling, the last-minute trade deal with the EU with its temporal caveats and policy blind spots, and the still inconsistent post-Brexit migration policy, it seems increasingly likely that Brexit will introduce new and more complex uncertainties with regards to the automation, or even the optimisation, of waged labour. Furthermore, a recent report by the Institute for the Future of Work (IFOW) demonstrated that a 1 per cent higher probability of automation was associated with a 1.4 per cent greater percentage of Leave votes (Gilbert and Thomas, 2021). The report suggests that places, such as Boston in Lincolnshire, Mansfield in Nottinghamshire, as well as Harlow in Essex and Great Yarmouth in Norfolk were at higher risk of job losses due to automation, on top of being Leave-voting constituencies. While in Boston up to 58 per cent of jobs were at risk of being negatively impacted by automation, that risk was negligible in places like the London boroughs of Wandsworth and Kensington and Chelsea, which supported Remain (Inman, 2019; Gilbert and Thomas, 2021).

COVID-19, essential workers and automation by stealth

If Brexit won't prove to be the prophesied catalyst of a smarter and technologically advanced economy, then maybe the COVID-19-induced social and economic disaster will be the tipping point? Beyond its undeniable and devastating impact on global public health, the COVID-19 pandemic has accelerated discussions about automating tasks. Robots have been deployed in hospitals to sanitise and sterilise, as well as to dispense medicines and deliver food (OECD, 2020). Beyond the health sector, efforts to improve customer support services (for example banking, utilities and so on) were boosted by the wider deployment of chatbots. Coombs (2020) has argued that the intensification of discussions around the faster adoption of automation and AI technologies seeks to compensate for the increased unavailability of human workers. Millions of people have lost their jobs, were made redundant or temporarily furloughed while millions of others have transitioned to working from home. Such rapid transformations have had a significant impact on budgeting and spending habits: high street shopping has all but disappeared at the expense of an unprecedented boom in online shopping where packing robots and low-paid workers in warehouses have ensured low prices. And while the spread of the pandemic has dealt a severe blow to jobs in hospitality and retail, it has also contributed towards the de facto demystification of automation. Specifically, there is evidence that consumer behaviour and preferences have changed in favour of automation and AI as more people became reliant on new communication technologies (Coombs, 2020).

Earlier aversion and distrust towards automation were often grounded within the idea that people prefer human interactions to interactions with machines. However, since COVID-19 made human contact exceptionally risky, avoiding it is increasingly perceived as a positive, which protects one's health and wellbeing. Self-service checkouts and a general preference for contactless payments have enabled social distancing and made the interaction between humans and machines acceptable and mundane (Coombs, 2020; Thomas, 2020). Further, the increased use of digital technologies, such as various platforms for video conferencing, has increased people's knowledge and technical skills and could create opportunities for the long-term deployment of such technologies (Howard and Borenstein, 2020). There is also evidence of an increased business confidence in automation and AI technologies: this is partly because robots cannot contract COVID-19, partly because the production process could be made safer if fewer people were to do repetitive tasks, such as cleaning supermarkets' and hospitals' floors at night, or screening patients' temperature (Coombs, 2020; Tucker, 2020).

Despite some of the successes of the technological changes outlined, it is worth noting that these remain patchy in terms of implementation, and unequal in terms of who gets to benefit from them. More importantly, currently we see no evidence of systematic planning and/or a consultative process at the government level in terms of a long-term sustainable rollout of automation. The elephant in the room here is that the COVID-19 pandemic 'has created economic impacts across the world far higher than the 2008–9 financial crisis' in that the levels of unemployment and, consequently, the availability of workers is substantial (Coombs, 2020: 3). The true scale of unemployment and labour availability is yet to become apparent, once the furlough scheme has been terminated and when Universal Credit payments are reduced to their usual levels.

If a coherent long-term automation strategy is not on the agenda, then would it be migrant workers who will provide high productivity and growth at less expense? Yes, most likely. In fact, the BEIS report discussed in this chapter concludes that despite empirical evidence that investing in automation technologies will increase productivity and growth in the mid- and long-term, there were examples where some UK employers articulated a preference towards manual labour done by migrant workers. For instance, car washes where low pay and exploitative working conditions make it cheaper for cars to be cleaned by hand than automated solutions (BEIS, 2019: 43–44). In other words, despite the potential of automation to incentivise businesses and employees towards more rewarding and less exploitative work, some UK employers expressed reluctance to invest in and adopt such technologies because, in the short term, they could achieve similar outcomes by relying on migrant workers.

Not unlike machines, migrant bodies are seen exclusively through the prism of their sheer functionality: as mere cogs, who have their basic human and labour rights gradually stripped away in order to appease post-Brexit anti-immigration attitudes. As the following paragraphs will demonstrate, seasonal migrants have been (ab)used in order to ensure the sustainability of supply chains with little to no regard for their health and safety and general wellbeing. Despite some short-lived praise that produce pickers are essential workers, post-Brexit migration policy governing low-skilled/low-paid mobility has degraded and disregarded them, thus effectively showing how 'essential workers' is a term for (lower) social class.

The Pick for Britain campaign exemplified lingering issues with post-Brexit migration policies uncertainties: regulatory provisions for seasonal mobility remained updated on an ad hoc basis despite continuous appeal from the agricultural sector for more certainty and predictability over the recruitment and retainment of seasonal workers. It was also a staunch reminder that the Brexit-related emancipatory appeal of investing in robots that would assist in the picking of soft fruit has not been backed up by investment and commitment. The fear that produce will be left to rot in the fields, which will have a knock-on effect on supply chains and consumption patterns, led to an almost paradoxical point of equilibrium where the UK government and recruitment charities (for example HOPS, Concordia) hatched plans to charter flights which were to bring in tens of thousands of workers – mostly from Bulgaria, Romania and Lithuania (Gallardo, 2021).

While COVID-19 has objectively made seasonal labour mobility more difficult, it also revealed a significant dependency on migrant labour. To be clear, this dependency is not exclusive to the agricultural sector in the UK – chartered flights were headed to Southern France, as well as places like Karlsruhe and Düsseldorf in Germany, along with Essex and the Midlands in the UK amidst fears that just-in-time supply chains cannot keep up with demand during national lockdowns. However, the dependency on migrant labour within the agricultural sector is especially acute in the UK as over 90 per cent of seasonal workers come from Eastern EU member states, the vast majority from Bulgaria and Romania. Political and public anxieties over whether and when migrant workers would arrive were amplified by chronic policy uncertainties plaguing the post-Brexit realm. Farmers and growers warned there was a real risk that thousands of tons of produce might be left to rot unless tens of thousands of migrants were flown to British farms. In the UK, up to 90,000 temporary positions had to be filled within weeks. Major agricultural companies and charities chartered flights to urgently bring in tens of thousands of seasonal agricultural workers, violating Eastern EU countries' national lockdowns in doing so.

Urgently flying in thousands of seasonal workers amidst a pan-European regime of lockdowns and travel bans is important for the following reasons. First, it effectively undermined the efforts of some Eastern EU governments in controlling the spread of COVID-19. In the earlier days of the pandemic, national governments in Eastern Europe were proactive in introducing lockdowns, curfews and travel restrictions. Such steps were necessary because in countries like Bulgaria and Romania there are high levels of emigration, combined with fast-growing ageing populations and unequal distribution of medical staff and facilities (Pencheva, 2020a). Second, it demonstrated an almost total disregard for health and safety regulations. Despite promises of social distancing on planes, photos of overcrowded airport lounges and bus stations in Romania emerged (*Euronews*, 2020). Photos and live footage showed no crowd control, no queues, face masks were worn by few, and hand sanitiser dispensers were nowhere to be seen. The closest commitment to health and safety was the promise/requirement that flown-in workers would undergo a 14-day mandatory quarantine upon arrival at their designated farms. This period of self-isolation, however, was to be unpaid. Further, analyses have demonstrated how seasonal migrants live four to a caravan, lack proper facilities and the most basic of workers' rights (Salyga, 2021a, b).

Third, for the most part, the British public has long been disassociated from the realities of low-paid manual labour and has grown accustomed to fresh and inexpensive produce produced by a disposable army of migrant workers (Pencheva, 2020a). Any public concerns about public health and the risks of community transmission of COVID-19 were dispersed by reiterating the fact that most UK farms are fairly remotely located, and most essential activities could be carried out without workers having to commute to nearby towns and villages. The chartering of flights to Eastern European capitals brought a sense of normality for many – the normality of being able to purchase affordable produce amidst an atmosphere of panic-buying and hoarding food. It also temporarily tempered the government's anti-immigration rhetoric, as well as negative public attitudes towards 'uncontrollable' migration from the rest of the EU. This is not to say that xenophobic attitudes and behaviours disappeared – on the contrary, there were reports of attacks on seasonal migrants in Italy (Matranga, 2020); rather, these were tempered by the realisation that there isn't a viable working alternative to cheap migrant labour.

The expensive and unethical efforts to urgently bring in migrant workers notwithstanding, there was still a significant gap in the British workforce. The Pick for Britain campaign was to address this gap on the practical level – by temporarily employing those who were disadvantaged by the negative economic impact of the pandemic, at the same time as providing a sense of patriotic duty fulfilment. The response to the campaign's appeal is perhaps best described as lukewarm and confused. After an initial spike in searches, which caused the recruitment website to temporarily freeze, not as many people as expected and needed applied for the advertised vacancies. In the peak of public interest, there were only around 10,000 applicants with even fewer having accepted work contracts due to low pay and demanding terms (O'Carroll, 2020). Official data showing the total number of job applications and the number of those who were offered and accepted contracts is currently unavailable due to the deregularised nature of the recruitment process, as well as the sensitive nature of some of the data. Speculations about numbers have also resulted in anecdotal evidence that British applicants were overlooked or ignored because the reality of the work, coupled with the dire conditions of pay and on-site accommodation, were further away from the rustic paradise initially portrayed by the campaign. A newly unemployed festivals organiser from Bristol shared his frustration on the pages of the *Guardian*:

'I live with my fiance and to live on site would mean I would only have one day a week for friends and family. They also said you can't use your own vehicle, which makes getting out to the shops difficult. Very quickly the romance of going to work for a farm to help provide food for the nation has become very unattractive,' he said. 'It seems it is very much geared up for migrant labour. We are not looking for special treatment, but the whole system needs to have some flexibility and not just have this blanket approach.' (Quoted in O'Carroll, 2020)

Such sentiments were echoed by other applicants who claimed that despite having applied for multiple jobs, they never got called for an interview. Public discussions were populated with people being frustrated about the untransparent recruitment process, as well as being labelled 'work-shy':

'I've applied for 200 jobs and you either get "We've got enough people now" or you don't hear anything back,' she said. 'Well, if they have jobs coming up at the end of May or June, why don't they just allocate those now and just confirm a job?' (Quoted in O'Carroll, 2020)

'I get an email from one farm worker recruitment agency saying I have not been shortlisted. How can this be? There seems to be a mismatch between demand for workers and matching workers to farm work.' (Quoted in O'Carroll, 2020)

On the one hand, one of the main takeaway points from the Pick for Britain campaign was that migrants have a better work ethic and higher levels of productivity compared to British workers. Some have estimated that migrant workers are nearly 50 per cent more productive than British workers (Goodwin, 2020). One farm-owner in Worcestershire told the *Sunday Times* that her British pickers – furloughed workers and students among them – were a 'very willing and enthusiastic bunch', but their productivity did not match the foreign workers they had replaced. 'If you're not used to working outside and you're not used to using your body all day, it is a real shock to the system for a lot of people', she said (Al-Othman, 2020). The call to undertake challenging and unglamorous jobs that are traditionally advertised to migrant workers problematises the use of the migrant body (see here Agamben, 2015) as a tool to achieve high levels of productivity and economic growth. The need for agile, strong and non-disabled bodies also goes to the heart of common understandings of 'skills' and further problematises the ways low-skilled labour is defined and measured. In this regard, it is also worth noting the contrast between how the same work was presented as a rustic paradise to British workers, while no regard was demonstrated for the safety, wellbeing and the very humanity of migrant workers.

On the other hand, the high levels of employability and productivity of Eastern European workers, as well as them being classed as key workers in high demand, didn't manifest favourably in the government's post-Brexit migration policy. The government's proposed points-based immigration system does not outline specific policy routes for low-skilled and/or temporary migration. Indeed, this type of migration is presented as the main problem that needs fixing. Some official guidance on sponsoring temporary and seasonal workers was published by the government in December 2020 and the Seasonal Agricultural Workers Scheme – branded a success – will be extended until the end of 2024. The pilot scheme tends to be renewed on an ad hoc basis, without firmer and clearer policy commitment. Indeed, guidance remains vague and insufficient despite sectoral lobbying for more provisions and flexibility. To make matters even more bizarre and complicated, in early 2021 the British government announced that the citizens of five countries – Bulgaria, Estonia, Lithuania, Romania and Slovenia – will not be eligible for reduced visa fees after Brexit (Gallardo, 2021). Coincidently, these nationals are overrepresented in the agricultural sector. The fiscal and political paradox of how politically problematic EU workers are, and how economically useful they are, continues to deepen and was further exemplified by the Pick for Britain campaign. In the absence of a long-term strategy for investing in automation technology, it seems that patchy, ad hoc and dehumanising post-Brexit migration policy will be the preferred option for achieving productivity and economic growth in the aftermath of Brexit and COVID-19.

Thus, a key consequence of the Pick for Britain campaign was that it brought together two subjectivities that are normally seen as distinct and separate: that of the migrant worker and that of the British worker. On the one hand, we have the subjectivity of the migrant worker, which embodied the notions of dispossession and economic utility. The COVID-19 pandemic was marked by the discursive shift from earlier claims that Eastern Europeans are stealing jobs towards a short-lived collective recognition of agricultural workers as key workers, along with NHS staff, refuse collection workers, delivery drivers and supermarket staff, among others. On the other hand, we have the dispossessed British worker – the new 'left behind' who were most severely impacted by the pandemic: those who were laid off/furloughed from hospitality jobs, students who didn't get to do their A-levels and/or who were thinking twice whether they can afford university education. Non-disabled but economically unproductive persons who had to be mobilised for the wider good of feeding the nation.

Moving forward

The overall argument this chapter has put forward is that considering the stage of automation in the UK, in the short and medium terms, employers at the low end of the labour market are most likely to continue to rely on migrant labour as a way of achieving high levels of productivity and growth at a lesser cost than if they were to invest in (some) automation technologies. Thus, the chapter has juxtaposed the abstract excitement about labour automation with the more tangible anxieties around migrant labour. The analysis has demonstrated that both migration and labour automation policies create conditions where cheap migrant labour is the preferred alternative to any tangible political commitment to automation in labour sectors dominated by migrant workers.

By critically engaging with the trope of *cheap labour* and the academic, political, media and policy debates around it, we have demonstrated how the perception of migrant workers as disposable cogs whose non-disabled bodies and work ethic are their only redeeming qualities has become discursively entrenched. However, the continued reliance on migrant workers, especially in the aftermath of Brexit and COVID-19, risks further diminishing the working rights of migrant workers. Although the problem at hand is undeniably complex – over-reliance on increasingly disenfranchised and dispossessed migrant workers versus a reluctance to invest in a more substantial labour automation rollout – it is imperative to note that this problem has serious policy implications.

For any post-Brexit immigration policy to be viable, it needs to recognise and overcome its internal paradox of conflating 'low-skilled' with 'low-waged' labour. The current system privileges earning potential at the expense of essential lower-paid jobs, which employers are currently struggling to fill. Examples here include the NHS, social care, hospitality, agriculture, food and drink processing. Furthermore, the current post-Brexit points-based system effectively creates a two-tier system whereby those who have a good command of the English language and earn £25,600 or more per year will have more rights and a clearer path to settlement than those who do not speak (good) English or cannot meet this salary threshold. The lack of clear guidance, policy predictability and a path to settlement with regards to migrant workers with lower earning potential could further increase the vulnerability of many essential workers. Seasonal and/or low-waged migrant workers will need to navigate a much more deregulated and precarious labour market in the aftermath of Brexit and COVID-19 and could easily fall prey to unscrupulous recruitment and employment practices. It is therefore imperative for current and future governments to prioritise building a more social and sensible post-Brexit immigration system, which offers better wages and working conditions to all workers, not just the highly qualified high earners.

Moreover, the lukewarm response to the Pick for Britain campaign, as well as its failure to provide a solution to the COVID-19-induced labour shortages in agriculture, should serve as a warning that achieving some level of automation is imperative for the sector moving forward. It is not a reasonable assumption that migrant workers should ensure the sustainability of supply chains, nor that unemployed/furloughed people should be galvanised under a patriotic banner to pick and sort produce, or risk being labelled 'work-shy'.

The analysis demonstrated that despite empirical evidence, as well as a Brexit-fuelled populist push to reduce the reliance on migrant workers, political will and commitment to invest in and to deploy automation technologies is simply lacking. We have pointed out that such a commitment would require not only a cross-party consensus but also a shift in the current partisan views on the role of the British state in managing labour market dynamics, as well as its ability to mitigate through progressive social policies the inevitable spike of intersectional inequalities that would follow from a more systematic rollout of automation technologies.

7

Nostalgia, Futurism and the Re-emergence of the Common

So, is it migrants or robots that steal our jobs? Our analysis has demonstrated, both theoretically and empirically, that neither is necessarily the case. Neither labour migration nor labour automation technologies 'steal jobs' as such; rather, we have argued that immigration and automation discourses are mutually constitutive and used to justify a divisive type of neoliberal governmentality, which weaponises public anxieties about job (in)security, self-realisation and welfare provisions. However, perceptions about stealing jobs persist because they are embedded within ideologically driven political rhetoric and policy choices, but also because both automation and migration policies are discussed in similar ways. There is the camp of the pessimists, which voices concerns over the availability/ scarcity of jobs, the levels of pay and the quality (or lack thereof) of work life. The optimists emphasise the opportunities that progressive migration/ automation policies present: more diverse and highly productive workforce, more leisure time because migrants/robots do the jobs that need doing but that no one wants to do.

Thus, this final chapter is not an orthodox concluding chapter in the sense that it does not seek to merely summarise our analytical findings. Rather, the current chapter tasks itself with moving beyond our main analytical findings and examines some political and theoretical alternatives to neoliberalism. In this chapter we critically examine and contrast the nostalgic approach of state interventionism and of homogenous traditional communities which became powerful campaign devices for the right-wing parties with the futuristic approach of Universal Basic Income (UBI), full *un*employment, and the constitution of communities independent of the structures and habits of work. By accepting the optimism of a future where work ceases to be a constitutive component for both individual and collective identity, we argue for a new configuration of the common that is able to be critical of itself through constant renewal and rejection of a social political order based on race and skills.

Communities of work

The demands for more automation and less immigration in conjunction with the normalisation of precarious working conditions have led to two social and political responses which problematise and occasionally transcend the established ideological boundaries of left and right, progressive and conservative. We name them the nostalgic approach and the futuristic approach respectively.

The nostalgic response mobilises nostalgia as a political force for critical examination and at times as an outright dismissal of present working conditions, welfare and social cohesion. Coined as a medical term in 1688 by Swiss physician Johannes Hofer, nostalgia has a wide analytical purchase: from neuroscience and clinical psychology to the discursive study of political rhetoric, ideology and propaganda. With regards to the latter, De Brigard (2020) argues that the concept of nostalgia is particularly useful because it does not require any real memories. In fact, nostalgia does not even need to be discursively embedded within objective historical realities, or even to be a part of living memory, which adds an important motivational aspect to it (De Brigard, 2020).

The nostalgic approach enters the political and social stage as soon as the discourses of automation and immigration are communicated by politicians, policymakers and campaigners as traumatic events and as processes capable of altering the fabric of everyday life and diminishing workers' status and integrity. Even though nostalgia and nostalgic sentiments can be diffused across the population, it is our intention to focus on a top-down understanding and examination of nostalgia. In particular, we are concerned with the ways nostalgia for a glorious past of job security, prosperity and cohesion has been manufactured by political parties, unions and movements for electoral gains. As we mentioned in Chapter 2, UKIP and the Brexit Party in the UK, Alternative für Deutschland in Germany, Lega in Italy and Rassemblement National in France reflect and weaponise the current dissonance between workers and the economy, the nation-state and globalisation, and communicate a return to a glorious past as the best political solution. Irrespective of their electoral success and political longevity, the deployment of nostalgia by these parties aims at restoring a sense of dignity and national pride to those who have been marginalised by the advent of automation and displaced by immigration.

The deployment of nostalgia in political discourse and policy should not only be seen as an experiment in political time-travelling and mere longing for an idyllic past of meaningful employment and strong social bonds but also as a process in which exclusion is practiced and legitimised. Yet, the nostalgic response is not exclusive to either established or pop-up far right and xenophobic parties. Crucial to the prominent position the nostalgic response occupies in policy and political communication are two different conceptualisations of nostalgia: an aggressive, uncompromising nostalgia dismissive of the present, and a managerial, compromising nostalgia that balances between progress and longing for the past. The former is closely associated with far right, anti-immigration, populist parties aiming at disrupting the political establishment and the latter has become an indispensable campaign and governing device for the so-called mainstream parties.

What unites populist and anti-immigration parties is the narrative of disappearing communities, broken social bonds and the gradual marginalisation of the White working class. Deindustrialisation, the unrealised promises of the knowledge economy and immigration have not only changed the structures of the labour market but more importantly the composition and meaning of class and belonging. The Brexit Party and the wider Vote Leave campaign brought to the foreground of British politics the victimisation of the post-industrial working class. Not so long ago class was considered a problematic and divisive issue in political and media discourses. Class had been replaced by references to social mobility and to Britain's hard-working families. The reluctance to talk about class divisions and inequalities was mainly a product of the Labour Party's strategy to position itself in the political centre and declare a truce among the working, middle and upper classes (Evans, 2000). New Labour's strategy was informed by the assumption as well as aspiration that the advent of knowledge economy would render the manual working class and class hierarchies obsolete.

The transition from Labour to New Labour legitimised the belief that old Labour was in terminal decline and that the working class was no longer relevant as an economic and political category (Crudas, 2008; Rutherford, 2011). New Labour's embrace of globalisation as an irreversible positive force inevitably determined the function of the state and government with regard to welfare and social protection: 'not to resist the force of globalisation, but to prepare for it, and to garner its vast potential benefits' (Blair, 2005).

The reappearance of the discourse of class at the beginning of the 2008 global financial crisis highlighted two rather incompatible manifestations of working-class culture. While class appeared to be a rather anachronistic, sociopolitical concept for the understanding of home ownership, taxation, workers' rights and wealth redistribution, the racialised class and more specifically the White working class became a potent explanatory tool for race and ethnic relations, the disappearance of traditional ways of life and for immigration partners and policies (Sveinsson, 2009).

The campaign of the Brexit Party in the 2019 European Parliament and general elections relied on the already established collective identity of the working class and communicated in dramatic terms its social marginalisation and cultural irrelevance in a world largely defined by the aesthetic values and economic principles of globalisation. As part of its European Parliament election campaign The Brexit Party (2019a) released a campaign video on Twitter titled 'Labour was the party of the working class'. The video opens with archive footage from the early and mid-1970s. Working-class Labour voters are watching on TV Labour MP Peter Shore giving his now famous talk in the Oxford Union debate in 1975. On the eve of the 1975 referendum on Britain's membership of what would become the EU, Peter Shore attacked the fear mongering of the 'pro-marketeers' and urged the nation to choose 'independence'. The Brexit Party video creates and at the same time capitalises on nostalgic sentiments with archival footage showing children cycling and playing in the streets and working men having a chat near a scrapyard.

The video establishes Labour's ideological break with its roots and traditional voters by projecting the dawn of New Labour in a completely different social and cultural context. Gordon Brown, Chancellor of the Exchequer from 1997 to 2007 and Prime Minister from 2007 up until 2010, is shown raising his ministerial box for the cameras outside 11 Downing Street followed by his now infamous exchange with the Labour voter Gillian Duffy whose concerns about immigration were dismissed as 'bigoted' (Weaver, 2010). The imagery of old Labour's heartlands, of Northern industrial towns populated by working men and children roaming free in the streets is replaced by the imagery of the City of London and of Islington in North London. In this new and shiny environment, children are wearing baseball caps, playing on their Xboxes and watching Tony Blair speeches on their MacBooks with their parents. A sequence of video extracts showing Lord Andrew Adonis on LBC radio insisting that Labour is the party of Remain, of Tony Blair at the World Economic Forum in Davos, and of bearded men tasting wine feeds into the perception that Labour are not only the party of Remain but also the party of the political and cultural establishment.

Soon after its success at the European Parliament elections, The Brexit Party (2019b) resumed the attack on Labour as part of its campaign to fill in candidates for the 2019 general election with a new video released on Twitter titled 'Labour is the party of Dalston, not Doncaster'. This time the Brexit Party focuses exclusively on the present and emphasises how neglected and poor Labour's heartlands have become. The video takes us to Doncaster, South Yorkshire. The town centre looks bleak, grey and deserted, shop fronts are padlocked, and roller blinds graffitied. Nigel Farage, the leader of the Brexit Party, is shown giving a speech in a local venue, telling his audience that "the Labour Party is now totally and utterly disconnected from its Northern roots, its people and its voters. Labour thinks you're stupid – the Labour Party thinks they know better. And it's the Labour Party now completely and utterly dominated from London" (The Brexit Party, 2019b). Nigel Farage uses the rhetorical technique of alliteration to emphasise Labour's betrayal of its heartlands. "Labour is a party that represents Islington not Islwyn, it's a party that represents Hampstead not Huddersfield, Dalston not Doncaster" (The Brexit Party, 2019b). White, elder men and women, traditional working-class people are shaking hands with Nigel Farage, affirming that the cultural void created by Labour has now been filled by the Brexit Party in a celebratory mood amplified by emotive music.

The visual material of the Brexit Party's campaign communicates a nostalgia for manual labour, Britain's industrial past and for closely knit working-class communities. Yet, the campaign does not extend to policies targeting automation of work, welfare or an overall industrial and post-industrial strategy. Instead, the Brexit Party's uncompromising nostalgia emphasises issues surrounding identity, belonging and tradition (Maronitis, 2021). In its attempt to discredit the present and by extension Britain's political establishment, the Brexit Party has successfully constructed an ethnically homogeneous group as victims of policies and cultural trends but in doing so disregards multiple forms of inequality and diverse experiences of poverty and insecurity. Such a nostalgia gains legitimacy through the making of the White working class and its subsequent association with Britain's former industrial regions, and on the other hand associates London with ethnic minorities and an abstract notion of the political and cultural elite.

The electoral success of the Brexit Party in the European Parliament elections and its ability to capture the public mood and frame the debate on working-class identity, immigration and Brexit during the 2019 general election pushed parties with governing experience and aspirations into an overdrive of nostalgia for traditional values and Britain's industrial past. Indicative of this managerial nostalgia that praises the past while acknowledging current achievements is the 2021 budget under the title: 'Build Back Better: Our Plan for Growth'. Prime Minister Boris Johnson, in his foreword to the budget, reproduces the all too familiar mantra of a harmonious working relationship between science and technology, the government and the private sector by emphasising that such a relationship has been central to the country's success since the Industrial Revolution and when industrial towns and cities generated sentiments of civic pride and aspiration. Similar to political and cultural observations and comments expressed before and during the elections of 2019, the prime minister draws attention to regional disparities and lack of opportunities outside London and the South East. Even though the UK is the fifth biggest economy in the world, the prime minister points out that there is a problem with the chronic 'distribution of opportunity' compared to the present 'distribution of talent'.

How does the government propose to solve the problem of regional disparities and bridge the gap between talent and distribution of opportunity? Unlike the 2017 budget, the answer is not to be found in automation, AI, autonomous vehicles and the total restructuring of the labour market, but in a programme of national renewal driven by unprecedented public spending. According to the Chancellor of the Exchequer, Rishi Sunak, this programme is based on the 'three pillars' of 'infrastructure', 'skills' and 'innovation'. Through a series of investment projects for the stimulation of economic activity and productivity, the government aspires to create new apprenticeships, raise the level of vocational skills and attract private investment in high growth companies and start-ups.

The renewed interest in public spending and the apparent disentanglement of governments from programmes of punitive austerity have raised questions over the relevance of the competitive free market. Does the government's nostalgia for a post-war Keynesianism indicate the end of the neoliberal logic? According to *The Economist* (2020b), a shift of this magnitude in economic and political thought only takes place once in a generation and the new approach taken by governments in Europe and North America confronts a new set of challenges. From the Keynesianism of the 1970s, to Friedman's monetarism in the 1980s and the independence of central banks in the 1990s, the approach outlined in the Build Back Better budget of 2021 will have to exploit the 'opportunities' and contain the 'risks' that state intervention generates (*The Economist*, 2020b). Despite the indication that a radical break has taken place, *The Economist* (2020b) insists that the exploitation of opportunities and the containment of risks must happen without a political takeover of the economy.

In order to assess the Build Back Better budget either as a nostalgic approach to the pressures for more automation and less immigration or as a novel approach that corresponds to unique problems caused by the COVID-19 pandemic, we have to look back at slogan heavily used by the UK government during the pandemic but first coined by the former President of the European Central Bank, Mario Draghi. Amidst the Eurozone crisis, Mario Draghi declared in June 2012 that the European Central Bank 'would do whatever it takes to preserve the euro. And believe me it will be enough' (quoted in Verdun, 2017; Tooze, 2018). The 'whatever it takes' statement communicates to citizens, markets and corporations that the state via its various agencies and institutions (such as central banks) can implement policies and take decisions previously inconceivable at least within the paradigm of the free market. Draghi's famous statement not only did preserve the euro but also created a convenient intellectual and political conundrum that allows the free market to be apolitical and at the same time to be able to attach itself to any political programme and ideology that is available.

Rishi Sunak's version of 'whatever it takes' oscillates between nostalgia and progress, necessity and determination. With the 'Build Back Better' budget, the Conservatives are justifying the expansion of the state and contextualising their own version of 'whatever it takes' by invoking the entrepreneurial and industrious spirit of the Victorians:

> Our mission is to unleash the potential of our whole country and restore the energy and confidence of the Victorians themselves. Just as the government has done whatever it takes to support lives and livelihoods throughout the Covid crisis, so we will turn the same ambition and resolve to the task of our recovery. (Rishi Sunak in HM Treasury, 2021)

Here, the reflective nostalgia for the Victorians indicates the ideological direction the expansion has taken and will continue to take under a Conservative government. By invoking an alleged golden era of imperialism and industry, the Conservatives denote the universal character and reach of 'Build Back Better' and more importantly indicate that such as project will not be restricted by concerns over welfare, equality and working conditions.

Communities off work

Parallel and in opposition to the nostalgic approaches of closely knit communities and expansive government spending, the idea of UBI has been promoted as the futuristic solution to the negative repercussions of the competitive labour market. UBI trials – if not in name, then at least in form, have been implemented in Europe (Finland, Netherlands, Spain), the US (temporarily from the late 1960s to the early 1970s for residents of New Jersey, Pennsylvania, Iowa, North Carolina, Seattle, Denver and Gary, Indiana; and more permanently for residents in Alaska), India, Kenya, Namibia, Brazil and Iran, among others (Samuel, 2020). Such cash support schemes tend to be temporally limited, funded by various sources (tax revenue, non-profits, non-governmental organisations and charities), and targeting unemployed people or people from socially disadvantaged backgrounds. Furthermore, the results from these trials are far from conclusive, which could hinder a more systematic policy implementation. On the one hand, participants reported an improvement in their wellbeing, as well as higher levels of trust towards their fellow citizens and national institutions and bureaucratic structures (Finland, Spain). In countries like India, Kenya, Namibia and Brazil, cash support reduced the levels of extreme poverty (Samuel, 2020). On the other hand, unemployed people were neither more, nor less, likely to be in employment following the cash support in the cases of Spain (Laín,

2019) and Finland (Young, 2019). In other words, the importance of work and work ethic – both for individuals' identities and in terms of generating economic growth and tax revenue, remains a key tenet behind UBI thinking (Finnish Ministry of Social Affairs and Health, 2016).

In Western liberal democracies, UBI remains more of an intellectual concept than a long-term, systematic public policy, as it is preoccupied with tackling inequality and the dysfunctional bureaucratic structures of welfare states. While the nostalgic approach has been unofficially adopted by conservative parties, the futuristic approach of UBI is predominantly a progressive concept and project that wishes to dissociate the left from state-centric economies and social organisation. UBI is being discussed as a futuristic approach in these pages mainly because it is or could be part of a future in which a guaranteed income is unconditionally distributed to all citizens. The main two objectives of UBI are the provision of a social and economic safety net from precarious working conditions and the elimination of poverty among people with low or no income at all. The apparent focus of UBI on individuals is mitigated by the aspiration of creating a more equal society in which employment and income cease to be the main determinants of inequality and exclusion. Besides this aspiration, UBI has entered mainstream debates because it constitutes a practical solution to a widely accepted problem: how will people be able to receive an income if most jobs are about to be automated?

While proponents of UBI insist that this is a politically neutral solution to the unchallenged advent of automation and by association to the problems of underemployment and precarity, we need to step back and examine the intellectual roots of this otherwise futuristic and challenging to implement idea. One of the earliest UBI advocates, Milton Friedman (2002), saw welfare programmes as a hindrance to the development of capitalism and the free market (see Chapter 2). Similar to many conservative thinkers and economists, Milton Friedman finds inspiration in the Victorians, when he argues that welfare should be part of private charitable initiatives. Not only does he attribute the gradual decline of private charity to the rise of welfare programmes across Europe and the US but more importantly he identifies the root of poverty in regulated capitalism. His solution to the hindrance of central planning and regulated capitalism is the introduction of a basic income that would eventually introduce all citizens to the price system and the principles of competition and at the same roll back the expansion of the welfare state. Friedman (2002) saw basic income as an effective substitution for a wide range of social welfare programmes such as pensions, education, public housing, health and social care. The desired outcome would be the incorporation of all these services into the market where citizens will be provided with choices that the basic income will afford them.

The redefinition of work and of the workspace in conjunction with increasing unemployment figures have created a fertile ground for the implementation of UBI. The reshaping of the labour landscape by the global COVID-19 pandemic has pushed governments to actions previously unimaginable and politically unacceptable. Although no government thus far has adopted UBI as an official policy, many have taken steps in this direction. From the UK's furlough scheme to the 'Trump Check' in the US, Germany's *Kurzarbeit* and France's Chomage Partial, governments during the pandemic have demonstrated that a guaranteed income for preventing the mass disappearance of jobs is not necessarily a far-fetched idea but rather a practical and, for the most part, socially acceptable solution. The possible implementation of UBI can either take the direction of Milton Friedman's doctrine of the primacy of the market and the price system or might contribute to the creation of new communities based on creativity instead of work and income. The latter direction informs the progressive case for the adoption of UBI. While conservatives think that UBI is a means to liberate individuals from the bureaucratic structures of the welfare state, progressives consider UBI as the first step towards the undermining of the competitive labour market. What unites these two different considerations is the reimagining and re-creation of communities in relation to capitalism.

Is it possible to formulate a progressive case for UBI that distances itself from the individualism of the price system and the free market? The main difference between Friedman's UBI and its progressive iteration can be detected in the scope and ambition of such a project. In the first instance, a progressive implementation of UBI as first outlined by Philippe van Parijs (2018) would help people meet all their basic needs without dismantling the welfare state. Yet, he argues that the implementation of UBI does not have to be a radical gesture posing an existential threat to the current structures of the capitalist economy. Philippe van Parijs and Yannick Vanderborght (2019) argue that UBI should be introduced as a moderate idea with the potential to transform the way people live and work. The political reintegration of an otherwise radical idea would require a rather limited understanding of universalism accompanied by specific conditions attached to it. The oscillation between moderation and transformation, universalism and particularism would make UBI accessible to citizens who are willing to undertake community work thus maintaining a sense of work ethic and preventing selective immigration flows.

Yet, the gradual integration of UBI into the capitalist economy is not a method shared by all UBI advocates. By placing UBI within a wider technological and political accelerationist framework, Srnicek and Williams assert that full unemployment as a desired and logical conclusion can only be supported by the introduction of UBI. The value of UBI, they add, should grow until up until the point when all goods and services can be purchased by this alternative wealth redistribution scheme. In their accelerationist framework, UBI would not only contribute to social equality and the disentanglement of life from work but would also accelerate the transition to a fully automated economy. Work in contemporary capitalism neither ensures wellbeing nor provides any sense of satisfaction and self-fulfilment. Here, the emphasis on the technological potentialities of capitalism is not about private profit but instead about social gains. This type of accelerationism, put forward by Srnicek and Williams (2015), associates the speeding up of techno-capitalism with radical social change. Such an association is theoretically underpinned by the tendency of capitalism to destroy itself and to communicate to its participants alternative social and economic futures. In this 'post-capitalist' vision the objective is not to save capitalism from the tensions and problems of automation, inequality and poverty but to eliminate all economic concerns and anxieties so people – the recipients of UBI – would pursue activities and form associations independent of need and waged labour. In such a society, the labour that remains will no longer be imposed upon us by an external force – by an employer or by the imperatives of survival. Work and other related activities will become driven by our own desires, instead of by demands from outside.

Theoretical debates on the origins, direction and implementation of UBI eventually reached political and policy discourses. Prior to the pandemic, in 2017, the Work and Pensions Committee rejected the very idea of a 'citizen's income' as a distraction from the real problems citizens and government departments faced, such as poverty, insecure work and budget cuts. For the Committee (2017) the two main problems surrounding the implementation of a citizen's income were practical and moral. Besides the perceived limited impact and appeal of the citizen's income, the Committee expressed concerns over funding and the inevitable rise in taxes and more importantly maintaining a sense of work ethic during turbulent times.

The impending apocalypse of mass unemployment communicated by think tanks and automation theorists (Chapter 3) and the inability to work (or go to work) because of the COVID-19 pandemic enabled MPs from Labour, the Liberal Democrats and the Green Party to reassert the need for a UBI (Hansard, 2020). Citing Finland and Utrecht in The Netherlands as successful case studies in which there were positive employment and wellbeing effects, UBI advocates argued that the complex and punishing welfare system in the UK should give way to a regulated system of cash transfers to all citizens. Despite the relative success of the implementation of a basic income in Finland and the Netherlands, and the urgent need to address the fear of rising unemployment figures, the Secretary of State for Work and Pensions, Thérèse Coffey, essentially recycled old arguments against unconditional cash payments: 'it is not targeted at the poorest in society and is not an appropriate way for us to try to distribute money' (Hansard, 2020).

The governmental and parliamentary response to UBI makes evident that dependency on wage labour and maintaining a work ethic based on precarity and fear are preferable to a fully automated economy in which unemployment is not a problem but rather a precondition for the constitution of new communities and associations. While UBI as a concept and a policy has the capacity to render the question who steals jobs irrelevant, its proponents have yet to confront the two great constitutive parts of the competitive labour market: national hierarchies and exclusion. Throughout the pages of this book, we explain that the discourse of automation is inexorably linked with the discourse of immigration. There is an explicit expectation that robots will have to work like immigrants and immigrants will have to work like robots. The accelerationist position relies heavily on the former while ignoring the latter. What are the limits of universalism when we consider the implementation of a UBI? Would the new communities and associations be independent of the current regimes of national exclusion and racial discrimination?

The common future

A few possibilities are being presented before us. On a political and administrative level, the nostalgia for homogeneous communities and for a strong, protective state appears to be ideologically and electorally successful. The state will protect jobs by celebrating the country's industrial past and at the same time restrict immigration by setting up salary thresholds and creating false distinctions between centre and periphery, globalism and localism, nationalism and internationalism. On an activist and academic level, the futurism of accelerationism and the imagining of a post-scarcity society ensures that there are no more jobs to be stolen. UBI will ensure the wellbeing of the unemployed and contribute towards the creation of communities independent of work and profit.

Notwithstanding the optimism that runs through the futuristic approach of accelerationism and UBI, we want to distance ourselves from the central position that the state could occupy in a fully automated economy and from a teleological understanding of technological progress. Considering its futuristic direction and the imagining of a world without work, the accelerationist proposition of UBI never quite manages to distance itself from the nostalgic response of the strong state and the revival of communities. A basic income issued unconditionally to all citizens will inevitably strengthen the state's role in administering and defining justice as well as governing the population. All struggles for justice, equality and welfare will have to be directed towards the state and in turn the state must maintain a political consensus among people whose only means of survival is UBI. Effectively UBI becomes a means of social control in a world devoid of political antagonisms. But the state is either incapable of addressing or unwilling to rectify social inequalities based on wealth and race. Throughout this book, we have seen how the state always aligns itself with the objectives of the competitive labour market, namely growth, productivity and labour flexibility. Our criticism of the role of the state is neither informed nor inspired by the neoliberal rejection of a state-centred economy but rather by the necessary cooperation between state and the free market for the sustainability and proliferation of a system built on competition, exclusion and precarity.

It is not only the close and problematic relationship UBI can develop with the state that concerns us here. In report after report, political statement after political statement, and policy after policy, technological progress and automation appear as unquestionable and irreversible processes. Accelerationism is in sync with the mainstream argument that automation needs to unroll uninterrupted in order to bring to an end the conundrum of low-paid jobs and low productivity. There would be no fear of having your job stolen because jobs, or meaningless, low-paid jobs, will no longer be required. It is difficult to disentangle the argument for a progressive acceleration from the capitalist future of a fully automated economy. Our concern in this book is not necessarily what will happen in the future but how automation is being used in the here and now for controlling the employed and the unemployed. To that effect, technology, or to be more accurate the anticipation of technological advancement, is neither apolitical nor neutral but instead a means to control and a site of struggle.

It is our contention, supported by the current policies on, and direction of, the competitive labour market, and the accompanying theoretical arguments on automation and immigration, that a new definition of the common is required for addressing the present inequalities between national citizens and immigrants, robots and workers. Before we proceed with the common and its application in world defined by the isolation and precarity of Homo Oeconomicus and its other, the Xeno Homo Oeconomicus, we need to evaluate the current revival of the notions of the commons and the common good. The revival we have in mind politicises ownership and acknowledges the importance of institutional spaces in which workers and citizens can act independently from the constraints and dictums of the market economy. Moving away from an understanding of the common as available and accessible resources to a space of sociopolitical action requires a critique of capitalism and its various manifestations in the spheres of work and communication. Unsurprisingly, the current revival of the common has been an integral component of the debate on the public use and value of digital platforms and capitalism.

The building of a 'digital commonwealth' requires an understanding of the powerful position universal digital platforms found themselves in relation to the state and existing regulatory frameworks. To that effect, Matthew Lawrence and Laurie Laybourn-Langton (2018), in an Institute of Public Policy Research report on the 'digital commonwealth', argue for more regulation concerning the reach and power of digital platforms in order to set up new standards for entering the market and handling data. Regulation, they argue, needs to be accompanied by the promotion of new platforms that are quintessentially smaller and operating under new regimes of ownership. Publicly owned and in collaboration with the universal platforms, these new local platforms could provide local services and assist citizens with more democratic tools for work and communication. The desired outcome of regulation of the existing universal platforms and the proliferation of small, publicly owned platforms would create, according to the authors of the report, a more 'mixed economy' that serves the common benefit and creates a 'digital commonwealth'.

One of the major problems we encounter when we engage with the binaries of the private and the public, and the private and the common, is our politically and historically distorted reference points. Far from resolving the differences between the state and the market, the national Homo Oeconomicus and the xeno Homo Oeconomicus, we have reached a point where terms such as private and public become meaningless slogans for the perpetuation of a system that thrives on confusion and complexity. The state gains political legitimacy when it protects private profit, and the market asserts its dominance when its presence and function are associated with the common good. Reports and theses on the constitution of the common good in and via a mixed economy overestimate our diminished collective capacity to distinguish between the private and the public and to theorise the common outside the confines of the market, precarity and survival (Lawrence et al, 2017; Labourn-Langton and Lawrence, 2018).

While the resurgence of the common good in political and economic debates (Standing, 2019) raises the issue of ownership, accessibility and availability, it ignores the more prominent question of power. Who has the power to define the common good and who precisely can build a political, economic and social framework in which the common good is shared and enacted? After all, restrictive immigration policies, precarious working conditions and low wages exist precisely because they contribute to the common good understood within the parameters of economic growth, national values and social cohesion. The focus on productivity and growth we highlighted in Chapter 2 either ignores inequality and exclusion or presents them as unavoidable yet regrettable manifestations of a competitive economy.

For the common good to make sense and be truly common we need to reassert its 'commonness' and set up a path where competition, individual performance and the fragmentation of Homo Oeconomicus along ethnic and national lines are no longer society's constitutive elements. The setting of such a path should not rely exclusively on state intervention or on what has been recently celebrated as 'the return of the state' during the COVID-19 pandemic. Following Dardot and Laval's (2018) theorisation of the common, we argue that the state is a false ally of the progressive politics of redistribution and equality. Our dismissal of the state is based on the notion that any policy, intervention or initiative including immigration controls and salary thresholds, levelling up and regional regeneration, training and reskilling for an automated economy does not aspire to contribute to the 'common good' but more accurately to mediate between different and conflicting social, political and economic interests.

The common and its various manifestations are not and should not be confined within a theoretical and conceptual framework. Instead, the common is theoretically informed and rooted in the reality of work, policies, fear and precarity. In Chapter 1, we argued that the experience of insecurity and flexibilisation of work are not limited to the working-class. Rather, we contextualised it as a process that affects both middle and working-class professions and in turn normalises precarity across class and ethnic categories. Should we try to find a common denominator for all workers and potential workers? Should our exploration of the common involve some kind of consensus among the national and xeno Homo Oeconomicus?

We want to maintain the optimism expressed in the futuristic approach to automation and employment while insisting on the constitution of a political community for articulating conceptions of work and immigration beyond the realm of competition and national sovereignty. The post-capitalist community cannot be possibly based on an ataraxia – a permanent political consensus manifested in the acceptance of certain working rights and social welfare provided and guaranteed by the state. Welfare in the form of 'levelling up' and 'build back better' only aspires to and occasionally succeeds in adapting the population to the demands of the market. The social protection, training and reskilling are provided in return for the rejection of any meaningful economic citizenship within the labour market and compliance with all the managerial norms and standards. Instead, the post-capitalist community envisioned here will be shaped by ruptures and continuous transformation. But we should not think that the interaction of workers in highly managed and controlled spaces leads to the creation and expression of alternative views and actions. If and when corporations allow or even encourage the mobilisation and interaction of workers, it is because the latter have been stripped of any political and economic power and reduced to mere operatives. In other words, the mobility of farm workers and the interaction between robots and workers in a highly supervised Amazon warehouse will neither undermine precarity as a means for governmental control nor will remove racial hierarchies in the labour market.

We must contest the call for a return to normality or even the promise of a bright, high-tech future that ignores the problems of the present. What is at stake is the necessity not only of including in the political and economic order those excluded according to race, class and education criteria but to create a collective subjectivity that critically corresponds to the introduction and implementation of automation and immigration policies that prioritise growth and productivity over rights and wellbeing. It would be the responsibility of this collective subjectivity to confront any hierarchies of race, class and gender by constantly inventing and communicating new political rights, new social and economic parts for those with no part in economy and society, and new spaces for mobility and expression.

Having a sense of meaningful collective control over the means of production – increased workers' control over ownership of workplace technology and higher levels of unionisation of the workforce (local and of migrant origin) in order to ensure adequate levels of collective bargaining – is the only way towards a more inclusive and equitable way of benefiting from future advances in technology and automation. Widespread engagement in such practices will ensure that the benefits of automation and technological advancement will not be disproportionately reaped by unelected and unaccountable corporate elites and shareholders at the expense of workers. Importantly, such bottom-up practices of social democracy have the potential to reshape and rewrite the social contract between the British state and its citizens and residents in a way that promotes a more sustainable and inclusive society.

References

6 River Systems (nd) Meet Chuck: a better way to fulfil. [online]. Available at: https://6river.com/meet-chuck/

Abizadeh, A. (2008) Democratic theory and border coercion: no right to unilaterally control your own borders. *Political Theory*, 36(1), 37–65.

Adkins, F. (2020) The fruitless saga of the UK's 'Pick for Britain' scheme. *Al Jazeera*, 19 November. [online]. Available at: www.aljazeera.com/featu res/2020/11/19/pick-for-britain-a-rather-fruitless

Agamben, G. (2015) *The Use of Bodies*. Stanford: Stanford University Press.

Aguis, C. (2010) Social constructivism. In A. Collins (ed) *Contemporary Security Studies*. Oxford: Oxford University Press, pp 49–69.

Al-Othman, H. (2020) Britain's fruit farmers hanker for return of foreign picker. *Sunday Times*, 23 August.

Alberti, G. and Danaj, S. (2017) Posting and agency work in British construction and hospitality: the role of regulation in differentiating the experiences of migrants. *The International Journal of Human Resource Management*, 28(21), 3065–3088.

Allen, W.L. (2016) A decade of immigration in the British press. Migration Observatory report, COMPAS, University of Oxford.

Alsos, K. and Eldring, L. (2008) Labour mobility and wage dumping: the case of Norway. *European Journal of Industrial Relations*, 14(4), 441–459.

Amable, B. and Palombarini, S. (2021) *The Last Neoliberal: Macron and the Origins of France's Political Crisis*. Translated by D. Broder. London: Verso.

Amazon Robotics (nd) Our vision. [online]. Available at: www.amazonr obotics.com/#/contact-us

Amazon Stories EU (2020) Thank you Amazon Teams. [online]. Available at: www.youtube.com/watch?v=PA6H46BjLmM

Anderson, B (1991) *Imagined Communities: Reflections on the Origins and Spread of Nationalism*. London: Verso.

Anderson, B. and Blinder, S. (2015) Who counts as a migrant? Definitions and their consequences. 4th revision. Migration Observatory briefing, University of Oxford.

Anderson, B. and Ruhs, M. (2012) Reliance on migrant labour: inevitability or policy choice? *The Journal of Poverty and Social Justice*, 20(1), 23–30.

Andrijasevic, R. and Sacchetto, D. (2016) From labour migration to labour mobility? The return of the multinational worker in Europe. *Transfer: European Review of Labour and Research*, 22(2), 219–231.

Arcarazo, D.A. and Martire, J. (2014) Trapped in the lobby: Europe's revolving doors and the other as Xenos. *European Law Review*, 39(3), 362–379.

Arcarazo, D.A. and Wiesbrock, A. (eds) (2015) *Global Migration: Old Assumptions, New Dynamics*, 3 volumes. Santa Barbara, CA and Denver, CO: ABC-CLIO.

Aristotle and Saunders, T.J. (1995) *Politics/Aristotle Books I and II/*. Oxford: Clarendon Press.

Asmus, R.D. (2008) Europe's eastern promise: rethinking NATO and EU enlargement. *Foreign Affairs*, 87(1), 95–106.

Balch, A. and Balabanova, E. (2011) A system in chaos? Knowledge and sense-making on immigration policy in public debates. *Media, Culture & Society*, 33(6), 885–904.

Balch, A. and Balabanova, E. (2016) Ethics, politics and migration: public debates on the free movement of Romanians and Bulgarians in the UK, 2006–2013. *Politics*, 36(1), 19–35.

Ban, C. (2016) *Ruling Ideas: How Global Neoliberalism Goes Local*. New York: Oxford University Press.

Barnard, C. (2009a) The UK and posted workers: the effect of Commission v Luxembourg on the territorial application of British labour law: case c-319/06 Commission v Luxembourg, judgment 19 June 2008. *Industrial Law Journal*, 38(1), 122–132.

Barnard, C. (2009b) 'British jobs for British workers': the Lindsey oil refinery dispute and the future of local labour clauses in an integrated EU market. *Industrial Law Journal*, 38(3), 245–277.

Becker, G.S. (1957) *The Economics of Discrimination*. Chicago: Chicago University Press.

BEIS (Business, Energy and Industrial Strategy Committee) (2019) *Automation and the Future of Work Report*, ordered by the House of Commons, 18 September. House of Commons: London.

Benanav, A. (2020) *Automation and the Future of Work*. London: Verso.

Berntsen, L. and Lillie, N. (2016) Hyper-mobile migrant workers and Dutch trade union representation strategies at the Eemshaven construction sites. *Economic and Industrial Democracy*, 37(1), 171–187.

Berriman, A. (2017) Will robots steal our jobs? The potential impact of automation on the UK and other major economies. In PricewaterhouseCoopers, *UK Economic Outlook 2017*. PricewaterhouseCoopers. [online]. Available at: www.pwc.co.uk/economic-services/ukeo/pwcukeo-section-4-aut omation-march-2017-v2.pdf

Beveridge, W. (1944) *Full Employment in a Free Society*. London: Allen & Unwin.

Black, R., Engbersen, G. and Okólski, M. (eds) (2010) *A Continent Moving West? EU Enlargement and Labour Migration from Central and Eastern Europe*. Amsterdam: Amsterdam University Press.

Blair, T. (2005) Conference speech, 2005. [online]. Available at: www.theg uardian.com/uk/2005/sep/27/labourconference.speeches

Blair, T. (2021) Without total change Labour will die. *New Statesman*, 11 May. [online]. Available at: www.newstatesman.com/politics/2021/05/ tony-blair-without-total-change-labour-will-die

Blanchflower, D.G., Saleheen, J. and Shadforth, C. (2007) The impact of the recent migration from Eastern Europe on the UK economy.

Blinder, C. and Richards, L. (2020) *UK Public Opinion toward Immigration: Overall Attitudes and Level of Concern*, 7th revision. Oxford: Migration Observatory.

Blinder, S. and Allen, W. (2016) UK public opinion toward immigration: overall attitudes and level of concern. Migration Observatory briefing, COMPAS, University of Oxford.

Boltanski, L. and Chiapello, E. (2007) *The New Spirit of Capitalism*. London: Verso.

Bonin, H., Eichhorst, W., Florman, C., Hansen, M.O., Skiold, L., Stuhler, J., Tatsiramos, K., Thomasen, H. and Zimmermann, K.F. (2008) Geographic mobility in the European Union: optimising its economic and social benefits. Research report no 19, Institute for the Study of Labor.

Borjas, G.J. (1999) Immigration and welfare magnets. *Journal of Labour Economics*, 17(4), 607–637.

Boswell, C. (2009) *The Political Usage of Expert Knowledge: Immigration Policy and Social Research*. Cambridge: Cambridge University Press.

Bourdieu, P. (1998) The essence of neoliberalism. *Le Monde Diplomatique*. [online]. Avalable at: https://mondediplo.com/1998/12/08bourdieu

The Brexit Party (2019a) Labour have now officially become a Remain party. *Twitter*, 10 July. [online]. Available at: https://twitter.com/brexitp arty_uk/status/1149009483600932864

The Brexit Party (2019b) Labour is the party of Dalston, not Doncaster. They have betrayed 5m of their voters. *Twitter*, 5 September. [online]. Available at: https://twitter.com/brexitparty_uk/status/1169553475790888960

Brown, W. (2015) *Undoing the Demos: Neoliberalism's Stealth Revolution*. Brooklyn: Zone Books.

Brown, W. (2019) *In the Ruins of Neoliberalism: The Rise of Anti-Democratic Politics in the West*. New York: Columbia University Press.

Buchowski, M. (2006) The specter of orientalism in Europe: from exotic other to stigmatized brother. *Anthropological Quarterly*, 79(3), 463–482.

Butler, S. (2017) Farmers deliver stark warning over access to EU seasonal workers. *The Guardian*, 21 February.

Buzan, B., Wæver, O. and De Wilde, J. (1998) *Security: A New Framework for Analysis*. London: Lynne Rienner Publishers.

Callison, W. and Manfredi, Z. (2019) Introduction: theorizing mutant neoliberalism. In W. Callison and Z. Manfredi (eds) *Mutant Neoliberalism: Market Rule and Political Rupture*. New York: Fordham University Press, pp 1–38.

Cameron, D. (2013) EU speech at Bloomberg. January. [online]. Available at: www.gov.uk/government/speeches/eu-speech-at-bloomberg

Carling, J. (2002) Migration in the age of involuntary immobility: theoretical reflections and Cape Verdean experiences. *Journal of Ethnic and Migration Studies*, 28(1), 5–42.

Carling, J. and Schewel, K. (2018) Revisiting aspiration and ability in international migration. *Journal of Ethnic and Migration Studies*, 44(6), 945–963.

Caro, E., Berntsen, L., Lillie, N. and Wagner, I. (2015) Posted migration and segregation in the European construction sector. *Journal of Ethnic and Migration Studies*, 41(10), 1600–1620.

Carrera, S., Eisele, K., Guild, E. and Parkin, J. (2015) The myth of benefit tourists and welfare magnets: a relationship between social welfare and free movement in the European Union? In D.A. Arcarazo and A. Wiesbrock (eds) *Global Migration: Old Assumptions, New Dynamics*. Santa Barbara, CA and Denver, CO: ABC-CLIO, vol 1, pp 251–271.

Carter, B., Harris, C. and Joshi, S. (1987) *The 1951–55 Conservative Government and the Racialisation of Black Immigration*. Coventry: Centre for Research in Ethnic Relations.

Castillo, D. and Ponce, A. (2020) Labour in the age of AI: why regulation is needed to protect workers. European Trade Union Institute. [online]. Available at: www.etui.org/Publications2/Foresight-briefs/Labour-in-the-age-of-AI-why-regulation-is-needed-to-protect-workers

Chamayou, G. (2021) *The Ungovernable Society: A Genealogy of Authoritarian Liberalism*. Cambridge: Polity.

Chandler, D. (2014) Beyond neoliberalism: resilience, the new art of governing complexity. *Resilience*, 2(1), 47–63.

Ciupijus, Z. (2011) Mobile central eastern Europeans in Britain: successful European Union citizens and disadvantaged labour migrants? *Work, Employment and Society*, 25(3), 540–550.

Clarke, S., Corlett, A., Finch, D., Gardiner, L., Henehan, K., Tomlinson, D. and Whittaker, M. (2017) *Are We Nearly There Yet? Spring Budget 2017 and the 15 Year Squeeze on Family and Public Finances*. London: Resolution Foundation.

Clayton, G. and Firth, G. (2018) *Immigration and Asylum Law*, 8th edition. Oxford: Oxford University Press.

Condliffe, J. (2017) Robots won't save the UK from a Brexit labor shortage. *Technology Review*, 21 December. [online]. Available at: www.technolog yreview.com/2017/12/21/146681/robots-wont-save-the-uk-from-a-bre xit-labor-shortage/

Conservative and Unionist Party (2005) Are you thinking what we're thinking? It's time for action. Conservative Party Manifesto. [online]. Available at: http://news.bbc.co.uk/1/shared/bsp/hi/pdfs/11_04_05_con servative_manifesto.pdf

Conservative and Unionist Party (2019) *The Conservative and Unionist Party Manifesto: Get Brexit Done: Unleash Britain's Potential.* London: The Conservative and Unionist Party.

Coombs, C. (2020) Will COVID-19 be the tipping point for the intelligent automation of work? A review of the debate and implications for research. *International Journal of Information Management*, 55, 102–182.

Copsey, N. and Haughton, T. (2014) Farewell Britannia? Issue capture and the politics of David Cameron's 2013 EU referendum pledge. *Journal of Common Market Studies*, 52(S1), 74–89.

Cremers, J., Dølvik, J.E. and Bosch, G. (2007) Posting of workers in the single market: attempts to prevent social dumping and regime competition in the EU. *Industrial Relations Journal*, 38(6), 524–541.

Crouch, C. (2011) *The Strange Non-Death of Neoliberalism.* Cambridge: Polity.

Crudas, J. (2008) A new politics of class: interview with Jonathan Rutherford. In *Soundings, 38: Cultures of Capitalism.* London: Lawrence and Wishart.

Daddow, O. (2012) The UK media and 'Europe': from permissive consensus to destructive dissent. *International Affairs*, 88(6), 1219–1236.

Daddow, O. (2015) Interpreting the outsider tradition in British European policy speeches from Thatcher to Cameron. *Journal of Common Market Studies*, 53(1), 71–88.

Dannreuther, R. (2007) *International Security: The Contemporary Agenda.* Cambridge: Polity.

Dardot, P. and Laval, C. (2013) *The New Way of the World: On Neoliberal Society.* Translated by G. Elliott. London: Verso.

Dardot, P. and Laval, C. (2018) *Common: On Revolution in the 21st Century.* London, New York, Oxford, New Delhi and Sydney: Bloomsbury.

Dardot, P. and Laval, C. (2019) *Never Ending Nightmare: The Neoliberal Assault on Democracy.* London: Verso.

De Brigard, F. (2020) Nostalgia Reimagined. *Aeon*, 20 July. [online]. Available at: https://aeon.co/essays/nostalgia-doesnt-need-real-memories-an-imagined-past-works-as-well?fbclid=IwAR1Fl9cefGNe1-h9377nsKF Ii53WgqvSJ6gkc1wpLTxtv2gs-BPKR6jZtvM

De Giorgi, G. and Pellizzari, M. (2009) Welfare migration in Europe. *Labour Economics*, 16(4), 353–363.

Delaney, K.J. (2017) The robot that takes your job should pay taxes, says Bill Gates. *Quartz*, 17 February. [online]. Available at: https://qz.com/911 968/bill-gates-the-robot-that-takes-your-job-should-pay-taxes/

Dellot, B. and Wallace-Stephens, F. (2017) The age of automation: artificial intelligence, robotics and the future of low-skilled work. RSA report, September.

Deloitte (2015) *From Brawn to Brains: The Impact of Technology on Jobs in the UK*. [online]. Available at: www2.deloitte.com/uk/en/pages/growth/articles/from-brawn-to-brains--the-impact-of-technol ogy-on-jobs-in-the-u.html#

Dennison, J. and Geddes, A. (2018) Brexit and the perils of 'Europeanised' migration. *Journal of European Public Policy*, 25(8), 1137–1153.

Department for Environment, Food and Rural Affairs (2020) Press release: up to 30000 workers to help reap harvest in 2021. [online]. Available at: www. gov.uk/government/news/up-to-30000-workers-to-help-reap-2021-harv est--2

Douglas, R.M. (2002) Anglo-Saxons and Attacotti: the racialization of Irishness in Britain between the world wars. *Ethnic and Racial Studies*, 25(1), 40–63.

Dustmann, C. and Frattini, T. (2014) The fiscal effects of immigration to the UK. *The Economic Journal*, 124(580), F593–F643.

Dustmann, C., Casanova, M., Fertig, M., Preston, I. and Schmidt, C.M. (2003) The impact of EU enlargement on migration flows. Home Office Online Report 25/03. [online]. Available at: https://discovery.ucl.ac.uk/id/eprint/14332/1/14332.pdf

Eatwell, R. and Goodwin, M. (2019) *National Populism: The Revolt Against Liberal Democracy*. London: Pelican.

The Economist (2020a) Covid-19 has forced a radical shift in working habits. *The Economist*. [online]. Available at: www.economist.com/briefing/2020/09/12/covid-19-has-forced-a-radical-shift-in-working-habits

The Economist (2020b) Free money: when government spending knows no limits. *The Economist*, 25–31 July, p 7.

Engels, F. (1845) Condition of the working class in England: Irish immigration. Marxists Internet Archive. [online]. Available at: www.marxi sts.org/archive/marx/works/1845/condition-working-class/ch06.htm

Esping-Andersen, G. (1999) *Social Foundations of Postindustrial Economies*. Oxford: Oxford University Press.

Euronews (2020) Skandal um Erntehelfer: Kein Abstand und Arbeiten in Quasi-Quarantäne. *Euronews*, 10 April.

European Commission (2011) Commission report on transitional arrangements regarding free movement of workers from Bulgaria and Romania. MEMO/11/773, 11 November.

European Monitoring Centre on Change (2005) *Knowledge Intensive Business Services: Trends and Scenarios*. Eurofound. [online]. Available at: www.eurofound.europa.eu/observatories/emcc/articles/working-conditions/knowledge-intensive-business-services-trends-and-scenarios

Evans, G. (2000) The working class and New Labour: a parting of ways? In R. Jowel, J. Curtice, A. Park, K. Thomson, L. Jarvis, C. Bromley and N. Stratford (eds) *British Social Attitudes: Focusing on Diversity – The 17th Report*. London: Sage, pp 51–70.

Evans, M. (2013) Migrants are given lessons in exploiting benefits; Bulgarians simply filling vacancies. *Telegraph*, 31 December, p 1.

Faist, T. (2013) The mobility turn: a new paradigm for the social sciences? *Ethnic and Racial Studies*, 36(11), 1637–1646.

Fassmann, H. and Munz, R. (1994) European east-west migration, 1945–1992. *International Migration Review*, 28(3), 520–538.

Feher, M. (2021) *Rated Agency: Investee Politics in a Speculative Age*. Princeton, NJ: Princeton University Press.

Ferguson, K. (2017) Low skilled EU workers will still be allowed to come to UK for 'seasonal' jobs post-Brexit, Gove says. *Daily Mail*, 2 October. [online]. Available at: www.dailymail.co.uk/news/article-4941378/Gove-Low-skilled-EU-workers-come-Uk-post-Brexit.html

Fernández-Reino, M. and Rienzo, C. (2021) Migrants in the UK labour market: an overview. Migration Observatory briefing, University of Oxford. [online]. Available at: https://migrationobservatory.ox.ac.uk/resources/briefings/migrants-in-the-uk-labour-market-an-overview/#kp1

Fernández-Reino, M. and Rienzo, C. (2022) Migrants in the UK labour market: an overview. COMPAS, University of Oxford, 6 January.

Fernández-Reino, M., Sumption, M. and Vargas-Silva, C. (2020) From low-skilled to key workers: the implications of emergencies for immigration policy. *Oxford Review of Economic Policy*, 36(1), S382–S396.

Finnish Ministry of Social Affairs and Health (2016) Legislative proposal on basic income experiment submitted to Parliament. Press release, 25 October. [online]. Available at: https://valtioneuvosto.fi/-/1271139/lakiehdotus-perustulokokeilusta-eduskunnan-kasiteltavaksi?languageId=en_US

Finotelli, C. and Sciortino, G. (2013) Through the gates of the fortress: European visa policies and the limits of immigration control. *Perspectives on European Politics and Society*, 14(1), 80–101.

Fitzgerald, D.S. (2019) *Refuge beyond Reach: How Rich Democracies Repel Asylum Seekers*. Oxford: Oxford University Press.

Fitzgerald, I. and Hardy, J. (2010) 'Thinking outside the box'? Trade union organizing strategies and Polish migrant workers in the United Kingdom. *British Journal of Industrial Relations*, 48(1), 131–150.

Fleming, P. (2017) *The Death of Homo Economicus: Work, Debt and the Myth of Endless Accumulation*. London: Pluto Press.

Flores Garrido, N. (2020) Precarity from a feminist perspective: a note on three elements for the political struggle. *Review of Radical Political Economics*, 52(3), 582–590.

Flynn, D. (2005) New borders, new management: the dilemmas of modern immigration policies. *Ethnic and Racial Studies*, 28(3), 463–490.

Forrester, K. (2017) Andrea Leadsom says young people should get excited about Brexit 'because robots will pick your raspberries'. *Huffington Post*, 3 October.

Foucault, M. (2004) *The Birth of Biopolitics: Lectures at the Collège de France*. Edited by M. Senellart and translated by G. Burchell. New York: Picador.

Foundational Economy Collective (2018) *Foundational Economy: The Infrastructure of Everyday Life*. Manchester: Manchester University Press.

Fox, J.E., Moroşanu, L. and Szilassy, E. (2012) The racialization of the new European migration to the UK. *Sociology*, 46(4), 680–695.

Freeman, C. (2006) Europe's 'Wild East' is ready to go west. Blighted by poverty and corruption, many in Bulgaria are eager to escape from their country when it joins the EU. But will the crime come too? *Sunday Telegraph*, 14 May, p 27.

Frey, C.B. (2019) *The Technology Trap: Capital, Labor, and Power in the Age of Automation*. Princeton, NJ: Princeton University Press.

Frey, C.B and Osborne, M.A. (2013) *The Future of Employment: How Susceptible are Jobs to Computerisation?* Oxford Martin School. [online]. Available at: www.oxfordmartin.ox.ac.uk/downloads/academic/The_Future_of_Employment.pdf

Friedman, M. (2002) *Capitalism and Freedom*. Chicago and London: Chicago University Press.

Future of Work Commission (2017) *Report of the Future of Work*. [online]. Available at: https://uploads-ssl.webflow.com/5f57d40eb1c2ef22d8a8c a7e/5f71a28418550428b77d4ab9_Future_of_Work_Commission_Report __December_2017.pdf

Gallardo, C. (2021) UK to discriminate between EU countries on fees for work visas. *Politico*, 4 February.

Gane, N. (2021) Nudge economics as libertarian paternalism. *Theory, Culture and Society*, 38(6), 119–142.

Garner, S. (2006) The uses of whiteness: what sociologists working on Europe can draw from US research on whiteness. *Sociology*, 40(2), 257–275.

Geddes, A. (2001) International migration and state sovereignty in an integrating Europe. *International Migration*, 39(6), 21–42.

Gertenbach, L. (2017) Economic order and political intervention: Michel Foucault on ordoliberalism and its governmental rationality. In T. Biebricher and F. Vogelmann (eds) *The Birth of Austerity: German Ordoliberalism and Contemporary Neoliberalism*. London and New York: Rowman & Littlefield, pp 239–261.

Geva, D. (2021) Orbán's ordonationalism as post-neoliberal hegemony. *Theory, Culture and Society*, 38(6), 71–93.

Gilbert, A. and Thomas, A. (2021) The Amazonian era: how algorithmic systems are eroding good work. Institute for the Future of Work (IFOW) report, May.

Glazer, N. and Moynihan, D.P. (1970) *Beyond the Melting Pot: The Negroes, Puerto Ricans, Jews, Italians, and Irish of New York City*, vol 63. Cambridge, MA: MIT Press.

Glenny, M. (1993) *The Rebirth of History: Eastern Europe in the Age of Democracy*. London: Penguin Books.

Goodwin, H. (2020) British fruit pickers 44% less productive than migrant workers. *The London Economic*, 23 August.

Goos, M., Arntz, M., Zierahn, U., Gregory, T., Gomez, S.C., Vazquez, I. and Gonzalez, J. (2019) *The Impact of Technological Innovation on the Future of Work* (Joint Research Centre). Seville: European Commission.

Gorz, A. (1982) *Farewell to the Working Class*. London: Pluto.

Gove, M. (2016) Interview for Sky News. 3 June. Available at: www.yout ube.com/watch?v=GGgiGtJk7MA

Government response to the BEIS Select Committee's twenty-third report (2020) 20 March. [online]. Available at: https://committees.parliament. uk/committee/365/business-energy-and-industrial-strategy-committee/ news/110717/beis-committee-publish-government-response/

Grant, W. (2017) Who will pick fruit and harvest vegetables after Brexit? Reviving SAWS could be a solution. *LSE Brexit*. [online]. Available at: https://blogs.lse.ac.uk/brexit/2017/11/17/who-will-pick-fruit-and-harvest-vegetables-after-brexit-reviving-saws-could-be-a-solution/

Gray, J. (2013) *Hayek on Liberty*. Abingdon: Routledge.

Guild, E. (2001) *Moving the Borders of Europe*. Leuven: Katholieke Universiteit.

Guild, E. (2002) The legal framework of EU migration: background paper. *The Political Economy of Migration in an Integrating Europe (PEMINT), Working Paper 2/2002*.

Guilluy, C. (2019) *Twilight of the Elites: Prosperity, the Periphery and the Future of France*. New Haven, CT: Yale University Press.

Hansard (2020) Hansard HC, Debate Pack, No. CDP0096, 8 October. [online]. Available at: www.parliament.uk/

Hansen, P. (2016) The definition of nudge and libertarian paternalism. *European Journal of Risk and Regulation*, 7(1), 155–174.

Hansen, R. (2000) *Citizenship and Immigration in Postwar Britain*. Oxford and New York: Oxford University Press.

Hardt, M. and Negri, T. (2001) *Empire*. Cambridge, MA: Harvard University Press.

Harvey, D. (2005) *A Brief History of Neoliberalism*. Oxford: Oxford University Press.

Hawkins, O. (2013) Migration statistics. House of Commons Library Standard Note SN/SG/6077. [online]. Available at: www.parliament.uk/briefing-papers/SN06077

Hayek, F.A. (1973) *Rules and Order*, volume 1 of *Law, Legislation and Liberty*. Chicago: University of Chicago Press.

Hayek, F.A. (2006a) *The Constitution of Liberty*. London and New York: Routledge.

Hayek, F.A. (2006b) *The Road to Serfdom*. London and New York: Routledge.

HM Treasury (2016) *Northern Powerhouse Strategy*. London: HMSO. [online]. Available at: https://assets.publishing.service.gov.uk/government/uploads/system/uploads/attachment_data/file/571562/NPH_strategy_web.pdf

HM Treasury (2017) *Autumn Budget 2017*. London: HMSO. [online]. Available at: https://assets.publishing.service.gov.uk/government/uploads/system/uploads/attachment_data/file/661480/autumn_budget_2017_web.pdf

HM Treasury (2020) *Budget 2020: Delivering on Our Promises to the British People*. London: HMSO. [online]. Available at: https://assets.publishing.service.gov.uk/government/uploads/system/uploads/attachment_data/file/871799/Budget_2020_Web_Accessible_Complete.pdf

HM Treasury (2021) *Build Back Better: Our Plan for Growth*. London: HMSO. [online]. Available at: https://assets.publishing.service.gov.uk/government/uploads/system/uploads/attachment_data/file/968403/PfG_Final_Web_Accessible_Version.pdf

Hobsbawm, E. (1998) The big picture: the death of neo-liberalism. *Marxism Today*, November/December 1998, pp 4–8.

Home Office (2006) *Points Based System: Making Migration Work for Britain*. [online]. Available at: www.official-documents.gov.uk/document/cm67/6741/6741.pdf

Home Office (2014) Review of the Migration Advisory Committee. [online]. Available at: https://assets.publishing.service.gov.uk/government/uploads/system/uploads/attachment_data/file/307172/TriennialReviewMAC.pdf

Home Office (2022) *EU Settlement Scheme Statistics*, updated on 13 January. [online]. Available at: www.gov.uk/government/collections/eu-settlement-scheme-statistics

House of Commons/Business, Energy and Industrial Strategy Committee (2019) *Automation and the Future of Work*. House of Commons. [online]. Available at: https://publications.parliament.uk/pa/cm201719/cmselect/cmbeis/1093/109302.htm

Howard, A. and Borenstein, J. (2020) AI, robots, and ethics in the age of COVID-19. [online] Available at: https://sloanreview.mit.edu/article/ai-robots-andethics-in-the-age-of-covid-19/

Hughes, L. and Daneshkhu, S. (2018) Gove calls for post-Brexit seasonal workers scheme for farms. *Financial Times*, 20 February. [online]. Available at: www.ft.com/content/787c7b44-163c-11e8-9376-4a6390addb44

Hugrée, C., Penissat, E. and Spire, A. (2020) *Social Class in Europe: New Inequalities in the Old World*. Translated by R. Gomme and E. Sanya Pelini. London and New York: Verso.

Huysmans, J. (2006) *The Politics of Insecurity: Fear, Migration and Asylum in the EU*. London and New York: Routledge.

Hylland Eriksen, T. (2007) *Globalization: The Key Concepts*. London: Berg.

Inman, P. (2019) Brexit voters more likely to live in areas at risk from rise of robots. *The Guardian*, 2 December.

International Transport Forum (2017) *Managing the Transition to Driverless Road Freight Transport*. [online]. Available at: www.itf-oecd.org/sites/default/files/docs/managing-transition-driverless-road-freight-transport.pdf

Isin, E.F. (2004) The neurotic citizen. *Citizenship Studies*, 8(3), 217–235.

Joppke, C. (1998) Why liberal states accept unwanted immigration. *World Politics*, 50(2), 266–293.

Jordan, D. (2020) Downing Street joins criticism of 'crass' job ad. *BBC*. [online]. Available at: www.bbc.co.uk/news/business-54505841

Joseph, J. (2013) Resilience as embedded neoliberalism: a governmentality approach. *Resilience*, 1(1), 38–52.

Kahanec, M. and Zimmermann, K.F. (eds) (2010) *EU Labor Markets after Post-enlargement Migration*. Berlin: Springer Verlag.

Kahanec, M., Zaiceva, A. and Zimmermann, K.F. (2010) Lessons from migration after EU enlargement. In M. Kahanec and K.F. Zimmermann (eds) *EU Labor Markets after Post-enlargement Migration*. Berlin: Springer Verlag, pp 3–46.

Kahanec, M., Zimmermann, K.F., Kurekova, L. and Biavaschi, C. (2013) Labour migration from EaP countries to the EU: assessment of costs and benefits and proposals for better labour market matching. IZA research report no 56.

Kelsen, H. and Schmitt, C. (2015) *The Guardian of the Constitution* (No. 12). Cambridge: Cambridge University Press.

Keynes, J.M. (2015 [1930]) Economic possibilities for our grandchildren. In R. Skidelsky (ed) *John Maynard Keynes: The Essential Keynes*. London: Penguin.

Kluge, A. and Negt, O. (2014) *History and Obstinacy*. Translated by R. Langston. New York: Zone Books.

Komlosy, A. (2018) *Work: The Last 1000 Years*. London: Verso.

Kopstein, J. (2009) 1989 as a lens for the communist past and post-communist future. *Contemporary European History*, 18(3), 289–302.

Kvist, J. (2004) Does EU enlargement start a race to the bottom? Strategic interaction among EU member states in social policy. *Journal of European Social Policy*, 14(3), 301–318.

Labour Party Manifesto (2005) *Britain Forward Not Back*. The Labour Party. [online]. Available at: http://news.bbc.co.uk/1/shared/bsp/hi/pdfs/13_04_05_labour_manifesto.pdf

Ladi, S. (2014) Austerity politics and administrative reform: the Eurozone crisis and its impact upon Greek public administration. *Comparative European Politics*, 12, 184–208.

Laín, B. (2019) *Report on the Preliminary Results of the B-MINCOME Project (2017–2018)*. Barcelona: Ajuntament de Barcelona.

Lawrence, M. and Laybourn-Langton, L. (2018) *The Digital Commonwealth: From Private Enclosure to Collective Benefit*. IPPR. [online]. Available at: www.ippr.org/research/publications/the-digital-commonwealth

Lawrence, M., Roberts, C. and King, L. (2017) *Managing Automation: Employment, Inequality and Ethics in the Digital Age*. IPPR. [online]. Available at: www.ippr.org/files/2018-01/cej-managing-automation-december2017.pdf

Leppard, D. (2013) Explosion in EU members forces ministers' hands. *The Times*, 15 December, p 11.

Lorey, I. (2015) *State of Insecurity: Government of the Precarious*. London: Verso.

MAC (Migration Advisory Committee) (2018) *EEA Migration in the UK: Final Report*. September. [online]. Available at: https://assets.publishing.service.gov.uk/government/uploads/system/uploads/attachment_data/file/741926/Final_EEA_report.PDF

Machiavelli, N. (2007) *The Essential Writings of Machiavelli*. Edited and translated by P. Constantine. New York: The Modern Library.

Mälksoo, M. (2006) From existential politics towards normal politics? The Baltic states in the enlarged Europe. *Security Dialogue*, 37(3), 275–297.

Mann, G. (2016) *In the Long Run We Are All Dead: Keynesianism, Political Economy and Revolution*. London: Verso.

Markova, E. and Black, R. (2008) The experiences of 'new' East European immigrants in the UK labour market. *Benefits*, 16(1), 19–31.

Maronitis, K. (2019) Robots and immigrants: employment, precarisation and the art of neoliberal governance. *Angles: New Perspectives on the Anglophone World*, 8. https://doi.org/10.4000/angles.570

Maronitis, K. (2021) The present is a foreign country: Brexit and the performance of victimhood. *British Politics*, 16, 239–253.

Marx, K. (1976 [1847]) The poverty of philosophy. In *Marx-Engels Collected Works*, vol 6. New York: International Publishers.

Marx, K. (1993 [1839]) *Grundrisse: Foundations of the Critique of Political Economy*. London: Penguin.

Marx, K. (2004 [1867]) *Capital: A Critique of Political Economy*, vol 1. London: Penguin.

Marx, K. and Engels, F. (1967[1848]) *The Communist Manifesto*. Translated by Samuel Moore. London: Penguin.

Matranga, A. (2020) Italian soldiers are enforcing a coronavirus barricade around community of migrant farm workers. *CBS News*, 26 June.

Mawby, R.C. and Gisby, W. (2009) Crime, media and moral panic in an expanding European Union. *The Howard Journal of Crime and Justice*, 48(1), 37–51.

Mazaranau, E. (2007) *Global Warehouse Automation Market Size 2012–2026*. Statista. [online]. Available at: www.statista.com/statistics/1094202/glo bal-warehouse-automation-market-size/

McDowell, L. (2009) Old and new European economic migrants: Whiteness and managed migration policies. *Journal of Ethnic and Migration Studies*, 35(1), 19–36.

Miller, D. (2010) Why immigration controls are not coercive: a reply to Arash Abizadeh. *Political Theory*, 38(1), 111–120.

Moravcsik, A. and Vachudova, M.A. (2003) National interests, state power, and EU enlargement. *East European Politics and Societies*, 17(1), 42–57.

Mortished, C. (2006) Minorities believe new migrants pose threat to jobs. *The Times*, 25 October, p 60.

Murgia, A. and Poggio, B. (2014) At risk of deskilling and trapped by passion: a picture of precarious highly educated workers in Italy, Spain and the United Kingdom. In L. Antonnuci, M. Hamilton and S. Roberts (eds) *Young People and Social Policy in Europe: Dealing with Risk, Inequality and Precarity in Times of Crisis*. Basingstoke: Palgrave Macmillan, pp 62–86.

Murji, K. and Solomos, J. (eds) (2005) *Racialization: Studies in Theory and Practice*. Oxford: Oxford University Press.

Nafziger, J.A. (1983) The general admission of aliens under international law. *American Journal of International Law*, 77(4), 804–847.

National Farmers' Union (2017) Drop in seasonal workers leaves some farmers critically short. *National Farmers' Union*. [online]. Available at: www. nfuonline.com/news/latest-news/drop-in-seasonal-workers-leaves-some-farms-critically-short/

Naudé, G. (2020) *Political Considerations upon Refined Politics, and the Master-stroke of State: As Practiced by the Ancients and Moderns*. London: Gale, ECCO.

Nowaczek, K. (2010) Pressure of migration on social protection systems in the enlarged EU. In R. Black, G. Engbersen, M. Okólski and C. Panţîru (eds) *A Continent Moving West? EU Enlargement and Labour Migration from Central and Eastern Europe*, Amsterdam: IMISCOE Research, Amsterdam University Press, pp 289–313.

O'Carroll, L. (2020) 'Just not true' we're too lazy for farm work, say frustrated UK applicants. *The Guardian*, 20 April.

Ochel, W. and Sinn, H.W. (2003) Social union, convergence and migration. *Journal of Common Market Studies*, 41(5), 869–896.

OECD (Organisation for Economic Co-operation and Development) (2020) *OECD Regions and Cities at a Glance 2020*. Paris: OECD Publishing. [online]. Available at: www.oecd-ilibrary.org/docserver/959d5ba0-en.pdf?expires=1615450425&id=id&accname=guest&checksum=863B4 93AEE765BCA6C5CA0941917B2CC

ONS (Office for National Statistics) (2017) Migration since the Brexit vote: what's changed in six charts. Office for National Statistics, 30 November. [online]. Available at: www.ons.gov.uk/peoplepopulationa ndcommunity/populationandmigration/internationalmigration/articles/ migrationsincethebrexitvotewhatschangedinsixcharts/2017-11-30

ONS (Office for National Statistics) (2020) Migration statistics quarterly report. Office for National Statistics, May. [online]. Available at: www. ons.gov.uk/peoplepopulationandcommunity/populationandmigration/int ernationalmigration/bulletins/migrationstatisticsquarterlyreport/may2020

ONS (Office for National Statistics) (2021) Long-term international migration, provisional: year ending December 2020. Office for National Statistics, 25 November.

Pencheva, D. (2019) Securitising Bulgarians and Romanians in British print media. Doctoral dissertation, University of Bristol.

Pencheva, D. (2020a) Coronavirus: flying in fruit pickers from countries in lockdown is dangerous for everyone. *The Conversation UK*, 21 April.

Pencheva, D. (2020b) Stealing jobs and benefits: Bulgarians and Romanians in British news media. In T.L. Thomsen (ed.) *Changes, Challenges and Advantages of Cross-Border Labour Mobility Within the EU*. Bern, Switzerland: Peter Lang Publishing, pp 28–62.

Pencheva, D. (2021) Detoxifying European migration (again). *Renewal: A Journal of Social Democracy*, 29(4), 75–83.

Pencheva, D. and Maronitis, K. (2018) Fetishizing sovereignty in the remain and leave campaigns. *European Politics and Society*, 19(5), 526–539.

Petre, J. and Walters, S. (2013) Exposed: the true cost of our open borders. *Mail on Sunday*, 29 December.

Pick for Britain (2020) #FeedtheNation. [online]. Available at: https:// feedthenation.co.uk

Pierson, P. (2001) Post-industrial pressures on the mature welfare state. In P. Pierson (ed) *The New Politics of the Welfare State*. Oxford: Oxford University Press, pp 80–105.

Polanyi, K. (1944) *The Great Transformation: The Political and Economic Origins of Our Time*. New York: Rinehart.

Prince, R. (2013) No skill, no entry. *Daily Mirror*, 23 October, p 26.

Recchi, E. and Triandafyllidou, A. (2010) Crossing over, heading west and south: mobility, citizenship, and employment in the enlarged Europe. In G. Menz and A. Caviedes (eds) *Labour Migration in Europe*. London: Palgrave Macmillan, pp 127–149.

Roberts, C., Parks, H., Staham, R. and Rankin, L. (2019) *The Future is Ours: Women, Automation and Equality in the Digital Age*. London: IPPR. [online]. Available at: www.ippr.org/files/2019-07/the-future-is-ours-women-automation-equality-july19.pdf

Robinson, D. (2013) Migration policy under the coalition government. *People Place and Policy Online*, 7(2), 73–81.

Rolfe, H. (2016) Employers' responses to Brexit: the perspective of employers in low skilled sectors. National Institute of Economic and Social Research. [online]. Available at: www.niesr.ac.uk/sites/default/files/publications/Employers%20and%20Brexit%20final.pdf

Rolfe, H. (2017) It's all about the flex: preference, flexibility and power in the employment of EU migrants in low-skilled sectors. *Social Policy and Society*, 16(4), 623–634.

Rolfe, H., Fic, T., Lalani, M., Roman, M., Prohaska, M. and Doudeva, L. (2013) *Potential Impacts on the UK of Future Migration from Bulgaria and Romania*. London: National Institute of Economic and Social Research.

Rose, N. (1993) Government, authority and expertise in advanced liberalism. *Economy and Society*, 22(3), 283–299.

Rousseau, J.-J. (1991) *The Essential Rousseau: The Social Contract; Discourse on the Origin of Inequality; Discourse on the Arts and Sciences; the Creed of a Savoyard Priest*. London: Penguin.

Rumelili, B. (2004) Constructing identity and relating to difference: understanding the EU's mode of differentiation. *Review of International Studies*, 30(1), 27–47.

Russell, H. (2020) 5am starts, poverty wages and no running water: the grim reality of 'picking for Britain'. *Prospect*, 30 May. [online]. Available at: www.prospectmagazine.co.uk/society-and-culture/pick-for-britain-advert-how-to-apply-pay-uk-fruit-and-veg-picking-jobs

Rutherford, J. (2011) The future is Conservative. In M. Glasman, J. Rutherford, M. Stears and St. White (eds) *The Labour Tradition and the Politics of Paradox: Compass, Fabian Society, Progress, Soundings*. London: Lawrence Wishart. [online]. Available at: www.lwbooks.co.uk/sites/default/files/free- book/Labour_tradition_and_the_politics_of_paradox.pdf

Samuel, S. (2020) Everywhere basic income has been tried, in one map. *Vox*, 20 October. [online]. Available at: www.vox.com/future-perfect/2020/2/19/21112570/universal-basic-income-ubi-map

Sanchez, S.I. (2015) Borders as floodgates: contesting the myth from federal and regional international experiences in light of EU free movement. In D. Arcarazo and A. Wiesbrock (eds) *Global Migration: Old Assumptions, New Dynamics*. Santa Barbara, CA and Denver, CO: ABC-CLIO, vol 1, pp 81–101.

Salyga, J. (2021a) Why migrant farm workers are living four to a caravan in a time of social distancing. *Jacobin Magazine*, 5 February.

Salyga, J. (2021b) In Britain, seasonal farm laborers toil for subminimum-wage piece rates. *Jacobin Magazine*, 6 April.

Savage, M., Cunningham, N., Devine, F., Friedman, S., Laurison, D., McKenzie, L., Miles, A., Snee, H. and Wakeling, P. (2015) *Social Class in the 21st Century*. London: Pelican.

Sayad, A. (2007) *The Suffering of the Immigrant*. Cambridge: Polity Press.

Schmidt, J. (2015) Intuitively neoliberal? Towards a critical understanding of resilience governance. *European Journal of International Relations*, 21(2), 402–426.

Schumpeter, J. (2010) *Capitalism, Socialism, and Democracy*. London and New York: Routledge.

Schumpeter, J. (2021) *The Theory of Economic Development*. Abingdon: Routledge.

Sivanandan, A. (2001) Poverty is the new black. *Race & Class*, 43(2), 1–5.

Skeggs, B. (2004) *Class, Self, Culture*. London: Psychology Press.

Slack, J. and Barrow, B. (2013) Domino's boss blasts Britons' work ethic. *Daily Mail*, 11 December.

Smith, J.E. (2021) *Smart Machines and Service Work: Automation on an Age of Stagnation*. London: Reaktion.

Somerville, W. (2007) *Immigration under New Labour*. Bristol: Policy Press.

Spencer, D. and Slater, G. (2020) No automation please, we're British: technology and the prospects for work. *Cambridge Journal of Regions, Economy and Society*, 13(1), 117–134.

Srnicek, N. and Williams, A. (2015) *Inventing the Future: Postcapitalism and a World without Work*. London: Verso.

Standing, G. (2019) *Plunder of the Commons: A Manifesto for Sharing Public Wealth*. London: Pelican.

Streeck, W. (2016) *How Will Capitalism End? Essays on a Failing System*. London: Verso.

Sumption, M. (2021) Briefing: work visas and migrant workers in the UK. COMPAS, University of Oxford, 17 September.

Sumption, M. and Vargas-Silva, C. (2020) Briefing: net migration to the UK. COMPAS, University of Oxford, 29 July.

Sumption, M. and Walsh, P.W. (2022) EU migration to and from the UK. COMPAS, University of Oxford, 15 February.

Sveinsson, K.P. (ed) (2009) *Who Cares About the Working Class?* London: The Runnymede Trust.

Terriff, T., Croft, S., James, L. and Morgan, P. (2000) *Security Studies Today.* Cambridge: Polity.

Tett, G. (2020) Bankers crave return of in-person trading floors. *The Financial Times*, 3 September. [online]. Available at: www.ft.com/content/ae63a 623-f367-4e69-adb6-efd11d7b65a9

Thomas, Z. (2020) Coronavirus: will Covid-19 speed up the use of robots to replace human workers? *BBC News*, 19 April.

Thompson, H. (2017) Inevitability and contingency: the political economy of Brexit. *The British Journal of Politics and International Relations*, 19(3), 434–449.

'Today Programme' (2017) *BBC Radio 4*, 23 November.

Tooze, A. (2018) *Crashed: How a Decade of Financial Crises Changed the World.* London: Penguin.

Touraine, A. (2000) *Can We Live Together? Equality and Difference.* Stanford: Stanford University Press.

Touraine, A. (2010) Sociology upside down: from systems to subjects. *New Cultural Frontiers*, 1(1), 4–15.

Travis, A. (2006) New limits planned on access to skilled jobs. *The Guardian*, 30 November, p 11.

Tronti, M. (2019) *Workers and Capital.* London: Verso.

Tucker, I. (2020) The five: robots helping to tackle coronavirus. *The Guardian*, 31 May.

Ulceluse, M. and Bender, F. (2022) Two-tier EU citizenship: disposable Eastern European workers during the Covid-19 pandemic. *Organisation*, 29(3), 449–459.

Van Parijs, P. (2018) Basic income and social democracy. In P. Van Parijs (ed) *Basic Income and the Left: A European Debate.* London: Social Europe Editions, pp 12–21.

Van Parijs, P. and Vanderborght, Y. (2019) *Basic Income: A Radical Proposal for a Free Society and a Sane Economy.* Cambridge, MA and London: Harvard University Press.

Vargas-Silva, C. and Walsh W.P. (2020) Briefing: EU migration to and from the UK. COMPAS, University of Oxford, 2 October.

Vauchez, A. and France, P. (2020) *The Neoliberal Republic: Corporate Lawyers, Statecraft, and the Making of Public-Private France.* Ithaca, NY and London: Cornell University Press.

Verdun, A. (2017) Political leadership of the European Central Bank. *Journal of European Integration*, 39(2), 207–221.

Vicol, D.O. and Allen, W. (2014) Bulgarians and Romanians in the British national press, 1 December 2012–1 December 2013. Migration Observatory report, COMPAS, University of Oxford.

Virno, P. (2003) *A Grammar of the Multitude*. London: Semiotext(e).

von Mises, L. (1998) *Human Action: A Treatise on Economics*. Alabama: Ludwig von Mises Institute.

Wacquant, L. (2009) *Punishing the Poor: The Neoliberal Government of Social Insecurity*. Durham, NC and London: Duke University Press.

Wæver, O., Buzan, B., Kelstrup, M. and Lemaitre, P. (1993) *Identity Migration and the New Security Agenda in Europe*. London: Pinter Publishers.

Walter, S. (2016) Crisis politics in Europe: why austerity is easier to implement in some countries than in others. *Comparative Political Studies*, 49(7), 841–873.

Weaver, M. (2010) The Gordon Brown and Gillian Duffy transcript. [online]. Available from: www.theguardian.com/politics/2010/apr/28/gordon-brown-gillian-duffy-transcript

Welfare Reform Act (2012) [online]. Available at: www.legislation.gov.uk/ukpga/2012/5/contents

Wieviorka, M. (2012) *Evil*. Cambridge: Polity Press.

Williams, R. (2020) *Culture and Materialism*. London: Verso.

Work and Pensions Committee (2017) *Citizens Income, Eleventh Report of Session 2016–17* (HC). 28 April, HC793.

World Economic Forum (2016) *The Future of Jobs: Employment, Skills and Workforce Strategy for the Fourth Industrial Revolution*. [online]. Available at: www3.weforum.org/docs/WEF_Future_of_Jobs.pdf

Young, C. (2019) *Experimental Finland*. OECD Observer. [online]. Available at: www.oecd-ilibrary.org/docserver/336216c4-en.pdf?expires=1650648 038&id=id&accname=guest&checksum=6FD2DE67464C9AD3C5804 724D8EE5D8F

Zebrowski, C. (2013) The nature of resilience. *Resilience*, 1(3), 159–173.

Zheng, R. (2018) Precarity is a feminist issue: gender and contingent labour in the academy. *Hypatia*, 33(2), 235–255.

Index